LIFE-SPANS

LIFE-SPANS
Or How Long Things Last

FRANK KENDIG *and*
RICHARD HUTTON

HOLT, RINEHART AND WINSTON
New York

First published in January 1980 by Holt, Rinehart and Winston,
383 Madison Avenue, New York, New York 10017.
Published simultaneously in Canada by Holt, Rinehart and
Winston of Canada, Limited.

Library of Congress Cataloging in Publication Data

Kendig, Frank.
    Life-spans, or, how long things last.

    Includes index.
    1. Life expectancy.   I.   Hutton, Richard,
joint author.   II.   Title.
QH528.5.K46   031'.02   79-15952

ISBN Hardbound: 0-03-053261-2
ISBN Paperback: 0-03-040876-8

First Edition

Designer: Betty Binns

Printed in the United States of America
10 9 8 7 6 5 4 3 2 1

Grateful acknowledgment is made to the following authors and
publishers to reprint excerpts from the following publications.
    "It's Alright, Ma (I'm Only Bleeding)." Words and music
by Bob Dylan © 1965, 1966 Warner Bros., Inc. All rights
reserved. Used by permission.

    The Man Who Walked Through Time by Colin Fletcher.
Copyright © 1967 by Colin Fletcher. Reprinted by permission of Alfred A. Knopf, Inc.

    Richard J. Harrison, William Montagna, *Man*, second edi-
tion, © 1973, p. 399. Reprinted by permission of Prentice-
Hall, Inc. Englewood Cliffs, N.J. 07632.

    The "How to Calculate Your Own Life-Span" section on pp.
26–28 of this volume is adapted from "Will You Live to be
100?" by Judith Bentley. Reprinted with permission *Family
Health* magazine, © January 1975. All rights reserved.

To D. G. and L. E.

The depressing thing is that this loudmouth bird is going to outlive me.
—owner of a white-naped
South American parrot

I've been selling life insurance for nearly 15 years and to this day not one of my clients has died. Actually one did die but he had not paid his premiums so I didn't count it.
—a New York City
insurance agent

Not marble, nor the gilded monuments
Of princes, shall outlive this powerful rhyme.
—William Shakespeare

CONTENTS

INTRODUCTION

T HE term *life-span*, at first glance, suggests biological life—specifically the span of time separating the birth and death of such animate organisms as people and pigs and petunias. As a practical or useful measurement, however, the notion of life-span has a much broader meaning. Astronomers, for example, are concerned with the birth and death of stars; anthropologists with the golden *age* of a culture; nutritionists with the shelf *life* of the foods we eat. And, of course, there was the man from Midas who used to guarantee his muffler "for the *life* of your car."

We are concerned here with the notion of life-span in its broadest sense. This is a book about how long things last—all sorts of things.

The longest life-span possible to imagine is that of the universe itself, already some 20 billion years old and still going strong. Most scientists, in fact, believe that the universe has an infinite life-span. It will simply go on existing forever. Even those scientists who forecast the universe's ultimate death agree that it has not yet reached middle age.

At the other end of the spectrum lies (or, more precisely, flits) a group of subatomic particles known as the unstable hadrons. These tiny particles exist for only one one-hundred-sextillionth of a second (10^{-23} second), less time than it takes for light to travel a single inch. Between these two awesome extremes fall the life-spans of everything else around us—from the North American continent to our pet goldfish, from the

Packard rusting in the backyard to the white line down the middle of the road.

Most of the life-spans presented in this book are expressed in terms of clock or calendar time. Neanderthal man had a life-span of only 29 years. The ostrich has a life-span of about 25 years. A lightning bolt has a life-span of about 50 microseconds. For many items, however, the incontrovertible measure of time does not best describe the idea of life-span. Consider, for example, the pencil.

Properly stored, a pencil could last thousands, perhaps millions, of years. But to what purpose? More helpful, in this case, is a life-span based on use. A hard pencil, it turns out, can write up to 30,000 words or draw a line more than 30 miles long. Similarly, the M-1 rifle, according to the U.S. Army, has a useful life-span of about 10,000 rounds of ammunition. And, of course, there is the new-car warranty—applicable for 12 months or 12,000 miles, whichever comes first.

The projected life-span of a baby born in the United States today is about 71 years, nearly double what it was at the end of the eighteenth century. This incredible rise in life expectancy has been caused largely by advances in medicine, public hygiene, and nutrition and is not likely to stop at 71 years. In fact, citing new breakthroughs in the study of aging, a few scientists go so far as to suggest that some of us alive today may never die.

Along with extending our own life-spans we have steadily increased the functional life-spans of the things around us. By establishing zoos and wildlife preserves we have extended the life-spans of many animals to longevities impossible for them to attain in the wild. By smoking, drying, salting, pickling, freezing, canning, and chemically treating the food we eat, we have been able to preserve it not only across seasons but even for decades. And by painting, lubricating, wrapping, and vacuum-packing our belongings, we have been able to greatly extend their useful lives. It could almost be said that the distinguishing characteristic of our curious species is its undying compulsion to preserve and maintain.

At present the state of the art of preserving inanimate objects is such that we can almost hand out immortality at will. At sufficiently low temperatures and pressures, for example, virtually everything—from human bodies to house paint—can be preserved indefinitely. Unfortunately, low temperatures and pressures are difficult and expensive to produce and maintain, so most of us are forced to make do with what we can. We pack our winter clothes in moth balls, stash our seldom-used belongings in the basement or attic, and relegate the rest to such storage spaces as the medicine chest, the cabinet under the kitchen sink, and the trunk of the family car—none of which provide ideal conditions to extend their life-spans.

We do have the advantage of one remarkable technological device, designed specifically for preservation—the refrigerator/freezer. It is a marvelous apparatus indeed, but it is also expensive, energy intensive, and often overcrowded. As one New Jersey woman put it, "I open the refrigerator and find my husband's golf balls, my son's photographic film, my daughter's cosmetics and no room for the leftovers."

There is no question that the pace of the so-called "space age" frequently makes our efforts at preserving the things around us seem fruitless. We live in what Alvin Toffler, author of *Future Shock*, calls a "throw-away" society. Diapers, Kleenex, paper napkins, plastic bags, no-deposit, no-return bottles, razor blades—the list of products made to be used once or twice and then discarded is virtually endless. It is now possible to buy paper clothing—worn once, then thrown away—as well as disposable toothbrushes complete with one brushing's worth of toothpaste.

Even products made to be durable seem to have shorter and shorter life-spans. As we have all heard, "they don't make them like they used to." Manufacturers argue that this shortening of product life is reasonable for, while manufacturing has become more and more automated, maintenance is primarily still performed by hand. As a result, it is often cheaper to replace than to repair. There was a time, for example, when a well-made pair of shoes could be worn for a decade or more with regular, reasonably inexpensive repair. Today, however, when new heels and soles for a pair of shoes can cost upwards of $20, it seems to be cheaper to buy poorly made shoes and simply throw them away when they wear out.

Despite what the manufacturers claim, many products are produced with limited life-spans simply so that more products can be sold. It is impossible, for example, to buy a necktie that has been stain-proofed, even though such protection could be furnished for only a few cents per tie. Similarly, cigarettes are manufactured with specially treated paper so that they will continue to burn even when they are not being smoked.

Interestingly, this situation seems to be changing as consumers are realizing that it is, in fact, *more* expensive to replace than repair. The change has been one of the few welcome consequences of the much-touted energy crisis. With energy prices soaring, it is becoming apparent that the final price of a product is determined not only by the cost of materials and labor but also by the cost of the energy required to produce it. One manufacturer of women's panty hose, with unwitting irony, has billed its product, Sheer Energy. As is frequently the case with advertisements, there is an element of truth here: These particular panty hose, it turns out, burn up the energy equivalent of a gallon of gasoline in the manufacture of each pair.

Whenever possible in this book we have tried to cite not only the average life-span of an object or organism but also hints or tips as to how its life-span might be extended. Also, if there is a known maximum or "record" life-span for a particular item we have tried to include it.

Our information was gathered from a vast spectrum of sources. The New York Public Library yielded many pieces of the puzzle. Especially valuable were such works as *The McGraw-Hill Encyclopedia of Science and Technology*, the *Co-*

lumbia Encyclopedia, *The Guinness Book of World Records*, the *CBS World Almanac*, the various libraries of Time-Life books, *The Woman's Day Encyclopedia of Cookery*, *The Doubleday Cookbook*, the *Encyclopedia of Fish Cookery* by A.J. McClane and his *McClane's New Standard Fishing Encyclopedia*, *Vanishing Birds* by Tim Halliday, and the charming *Mythological Creatures* by Paulita Sedgwick.

Special credit should be given to the Bell System's *Yellow Pages*, which saved each of us several pairs of shoes.

Many government agencies, universities, and consumer groups were instrumental in the preparation of *Life-Spans*. Special thanks are due to the Rutgers Food Study, the U.S. Department of Agriculture, as well as to professional and manufacturing associations of all kinds.

Finally, best wishes and unbridled gratitude to the following, without any one of whom this book would not have been possible: Douglas Gassner who, when all we were considering were the life-spans of living things said, "What about the white line in the middle of the road?" Don Hinkle who steadfastly badgered the public relations people of corporate America and even seduced the U.S. Army into "standing ready" to help us when we needed it. Penny Hinkle who not only can cook a beautiful meal but was able to find out how long it would last. Barbara Ford who has written more about animals than almost anybody alive and who took on the laborious chore of finding out their life-spans. Ann Bussard who opened up her medicine cabinet, then got on the telephone to find out the life-spans of everything in it. Douglas Colligan who set aside his own writing to take up the slack when it seemed that *Life-Spans* would never be finished. Jane Wisan who charted and ordered and listed the life-spans of foods without becoming *too* involved in the work. Lena Eriksson who dug through piles of consumer information and somehow never forgot how to laugh. David Bischoff who put an armlock on the U.S. Department of Transportation. Owen Davies who took a floundering manuscript and added the finishing touches, far beyond what he was asked to do. Don Hutter who thought enough of the idea to contract the book in the first place, and then stuck by us through some very rough moments. Barbara Lowenstein, our mutual agent, who prodded, argued, cajoled, and did everything she could to turn an idea into a book. Debby Gobert who stuck by one recalcitrant author, even married him, who kept the game going when it mattered most. Keri Christenfeld who went far beyond her duty as an editor, who rolled up her sleeves, got on the telephone, took out her pencil, and, perhaps more than anyone else, made sure that *Life-Spans* became a book.

LIFE - SPANS

1 / THE HUMAN LIFE-SPAN

THROUGH the 250 millennia or more of man's life on earth, the "expectable" span of his lifetime has risen—gradually over most of history, sharply over the last 100 years. To a large degree we can point out and even measure the causes of this increased longevity, this capacity of man's—aided by whatever cultural advantages or advances—to forestall death. What we cannot yet do is explain death itself or, ultimately, the aging process that makes death inevitable. And both death and the process of aging are as much a part of a life-span as life itself.

The controversy surrounding the issue of abortion has brought attention to another difficulty in evaluating man's life-span: determining the point at which life actually begins. Some claim that life exists at the moment of conception; others, at the moment when the fetus can sustain life independent of its mother; and still others, at the moment of birth. There is little question, however, that the process of *aging* begins with conception and that it ends with death. And before we can properly consider life-spans, we should try to understand just what aging is all about.

Probably the most misunderstood aspect of aging and death is old age—what it is and what it does. Simply speaking, old age is the latter stages of a constant, continuing, and inexorable decline in the ability of the body to carry out its own functional requirements and to resist and bounce back from disease and trauma. Even if we somehow escape the ravages of war, accidents, and deadly diseases, this decline proceeds at a steady pace, so that, for most of us, there is a good chance that we will succumb to physical ailments between the ages of 65 and 80. What this means is that there really is no such thing as death *by* old age. What the elderly are actually dying from is a combination of disease or disorder and the inability of their bodies to respond to the challenge any longer. As Alex Comfort noted in his book, *Aging: The Biology of Senescence*, if we kept throughout life the same resistance to stress, injury, and disease that we had at the age of 2, about half of us here today might expect to survive another 700 years.

Why can't we retain our youthfulness?

The medical profession has done much to combat the external causes of death that we face daily. It has learned to patch and sew up many traumatic injuries, fight dangerous germs with friendlier varieties, and even replace some of our bodily parts as they become defective. But it has barely begun to deal with the most fundamental causes of death—the general disintegration and ultimate breakdown of our bodies.

Today, a growing band of medical researchers is probing the nature of aging. These pioneers are actually seeking ways to slow the deterioration that overtakes us with time, to put an end to death itself. Theirs is a relatively new field and a small, ill-funded one compared with other branches of medical research. Yet there are signs that they may succeed long before most nonscientists would predict. As early as 1964, 82 gerontologists who replied to a questionnaire concluded that medicine could add 20 years to the human life-span by 1992. A second survey, in 1975, found that scientists and industrial planners expected anti-aging drugs to be developed by the year 2000.

So far, no one has figured out how aging occurs. So many bodily processes change with time that there is no sure way to tell which cause aging and which are its effects. Some may be caused by aging and may promote further deterioration as well. Sorting them out is slow and costly work, for test animals must live out their full lives before an experiment is over. In studies of rats, this can take up to 4 years; with larger animals, experiments take much longer. Despite these difficulties, theories abound.

According to one theory, organs and organ systems within our bodies age steadily throughout our lives, losing about 1 percent of their functioning capacity per year. As a result, by the time we are 70, much of our functional activity has already shut down, slowing our ability to produce vital materials like hormones and regulate such vital functions as blood pH and pulse rate.

Another hypothesis is that our tissues simply begin to wear out after a while. They have a limited ability to renew and replace themselves, a limited capacity to heal.

The current trend in research is not to deny the basic truth of such hypotheses, but to uncover what exactly is going on at the cellular level throughout the aging process and to determine those changes in body chemistry responsible for cellular deterioration. For example, many scientists believe that our genes carry some form of self-destruct mechanism that ends our lives at predetermined times. Dr. Leonard Hayflick is one scientist who subscribes to this *death clock* theory. Working first at Philadelphia's Wistar Institute and then at Stanford University, Dr. Hayflick has found that animal cells, cultured or grown in the laboratory, are able to divide only a limited number of times before they stop. All animals, including man, contain cells that continue to divide throughout their lives. But while cells taken from young animals and grown in the laboratory divide many times before their cultures die out, those from senile animals divide only once or twice. In man, cells can reproduce through about 50 cellular generations; then the line dies out. Can it be that we die because our cells can no longer reproduce? Dr. Hayflick is certain of it.

Other researchers reject both the notion of an internal clock that eventually shuts off the body's cells and the theory that chemical alterations in our tissues cannot be prevented. Instead, they believe that senescence is the result of random forces, that we age because some accidental, repetitive injury eventually destroys our vitality. Typical of these theories, and one of the most highly regarded, is the *error catastrophe hypothesis* proposed by Dr. Leslie Orgel of the Salk Institute in La Jolla, California. Dr. Orgel believes that in building its component proteins, the body occasionally makes a mistake and creates a defective protein. Whatever function that protein was to perform then goes unfulfilled. For example, an enzyme required for energy transport within the cell might be unable to do its job. This would cripple the cell and slightly impair the function of the organ containing it. Because some proteins aid in the reproduction of DNA—the genetic material that controls all life processes—defects in the protein would return as errors in the DNA. When too much damage accumulates, the system breaks down and the organism dies.

If this idea holds up, it offers a way to extend the human life-span. The cell contains enzymes whose sole purpose is to repair faulty genetic material. Synthetic DNA could be made with deliberate errors and used to stimulate production of the repair enzymes. There is no guarantee this will work, however. Though defective proteins do accumulate with age, they may well be an effect rather than a cause. Some mathematicians studying the problem say that too few genetic errors occur to account for the deterioration caused by senescence.

A second random-accident theory holds that protein molecules lose their biological activity because they become linked together. The theory is usually applied to collagen, the protein that keeps the skin and cartilage supple. The skin sags and joints become stiff, it is said, because the collagen becomes cross-linked. The process attacks other organs as well, eventually preventing them from functioning. Dr. Johann Bjorksten, who originated this *cross-linkage theory*, is so sure it is correct that he has claimed he can see no reason to test other possible mechanisms. Dr. Bjorksten's certainty is understandable. Working at the Bjorksten Research Foundation, in Madison,

Wisconsin, he has discovered bacterial enzymes able to dissolve the bonds that cross-link collagen in aging mice. Though side effects are severe, mice given the enzymes live longer.

Other researchers have suggested that other highly reactive chemicals called *free radicals* promote cross-linking and are possibly the cause of aging. They have found that drugs that destroy free radicals also extend the life-spans of test animals. Between them, these experiments provide strong proof that cross-linking is at least a contributor to the aging process.

Other gerontologists are far from convinced that Dr. Bjorksten's ideas really offer a complete explanation of why we age. For example, early researchers suspected a pigment called lipofuscin, which accumulates in aging animals, of being a kind of garbage that progressively clogs cells until they no longer function. At one time, this concept had almost been abandoned. It turns out, though, that mice also live longer when given drugs that slow the production of lipofuscin.

Experiments performed by Dr. W. Donner Denckla of the Gerontological Research Center in Baltimore have shown during aging the pituitary gland begins to release a chemical that interferes with metabolism throughout the body. Sufficient amounts of this decreasing-consumption-of-oxygen (DECO) hormone could cause virtually all the effects seen in aging. Dr. Denckla thinks it is the DECO hormone, or a similar one, that finally kills. So far, the endocrinologist has found only evidence of the hormone, and has not isolated the DECO hormone itself, but he is certain that the pituitary is the cause of aging. By removing the pituitary glands of aging mice and giving the mice hormone supplements to make up for the other secretions lost, he has temporarily rejuvenated the animals. Yet they die on schedule.

Dr. Norman Orentreich, a New York City dermatologist, has rejuvenated animals by removing the proteins from their blood plasma. It is among these proteins that Dr. Denckla's death hormone would be found. But Dr. Orentreich denies that his success means he is ridding the animals of a DECO hormone. The technique, called plasmapheresis, could operate by eliminating some toxic waste product or by a mechanism yet unknown. It does seem, at least, that whatever process accounts for the success of plasmapheresis, it must affect the entire body. Like Dr. Denckla's mice, Dr. Orentreich's rodents and dogs die after living out normal life-spans.

There are other theories of aging —almost as many as there are theorists. Most have evidence to support them, and many conflict with others whose proofs seem as good. If it is the DECO hormone that kills us, why do our cells eventually die out when grown in test tubes, where there is no hormone to destroy them? If internal clocks shut off the cells themselves, why does removing the pituitary glands or plasma proteins restore mice and dogs to adolescent health? And if aging is programmed in any fashion, why do mice live longer when given free-radical inhibitors of Dr. Bjorksten's bacterial enzymes?

There is a growing feeling among gerontologists that no one theory can explain everything we know about aging. We grow old because several different processes attack our bodily functions. Perhaps we age because cross-linking and the DECO hormone weaken us, and we die of ills we no longer have the strength to combat. The cellular death clock may be a backup mechanism that functions only when illness does not arrive on schedule. If so, it is a hopeful discovery for scientists working to extend human lives. If there are many aging processes, there will be many ways to promote longevity. At some future time our lives may be measured in centuries, not decades. Well within the life-spans of many who read this book, the life-expectancy formula that is now the best we have to offer will almost surely be left far behind. But until then, perhaps the most cogent comment upon our life-spans—and the life-spans of all living things—has come from Bob Dylan: "Those who are not busy being born are busy dying."

Having acknowledged these larger, more conjectural questions of human longevity—those aspects of it that may be *least* measurable—we can turn to the subject of life-spans and take it up, thanks to the modern science of statistical analysis, in terms we *can* measure.

We begin with a baby born in the United States today. It has an expected life-span of 71.3 years. But there are babies and there are babies. According to a 1974 study by the National Center for Health, a white baby born in the United States has an expected life-span of 72.7 years; a nonwhite baby, only 67 years. And, as is the case in virtually all species, females can expect to outlive males: A female white baby born in 1975 has a projected life-span of 77.2 years, whereas a male white baby can expect to live only 69.4 years. In some countries the difference is astounding: A woman's life expectancy in Gabon, Africa, for example, is 45 years; that of her spouse is only 25.

No matter what your category, your projected life-span increases as you get older. Statistically, you become a survivor. A woman who reaches 70, for example, has a projected life-span of 83.6 years. By the time she reaches 85, her projected life-span is up to 90.8 years. Curiously, the situation with respect to race and color seems to reverse with increasing age. Whereas a white female child can expect to outlive her nonwhite counterpart, a nonwhite woman over 70 can expect to outlive a white woman of the same age.

Most scientists feel that a human body's potential life-span has always been the same—about 100 years. There are, of course, reports of human beings living well over 100 years, and even some suggestion that humans have approached 200 years, but such claims have not held up to investigation. According to the *Guinness Book of World Records*, the longest authenticated life-span of a human being is 113 years, 214 days.

Historical Man

Certainly some of us may feel cheated that our actual projected life-spans in the advanced society of the United States are only a little over 72 years out of a possible 100. To make matters seem even worse, only 10 percent of the population actually survives past age 65; in 1973 there were only 542,379 American men (0.27 percent of the population) and 968,522 women (0.48 percent) 85 years old or older. Nevertheless, even this is an astonishing accomplishment; during more than nine-tenths of human history, the average life-span was less than half that. The following life-spans, based on studies of skeletons, legal documents, and gravestones, among other effects, cover human beings who lived during the first 90 percent of human history and who survived to at least 15 years of age.

	Life-span (years)
Neanderthal man	29
Cro-Magnon man	32
Man in the Copper Age	36
Man in the Bronze Age	38
Greek and Roman man	36
Fifth-century man (England)	30
Fourteenth-century man (England)	38
Seventeenth-century man (Europe)	51
Eighteenth-century man (Europe)	45*

*Epidemics in urban areas were the cause of this drop in life-span.

INFANT MORTALITY

If infants and children were included in these life-span figures, as they are in the determination of life expectancy today, the figures would be much lower because of the enormous toll of childhood diseases throughout most of history. According to Edward Shorter in *The Making of the Modern Family*, only two out of every three infants survived their first year, and half of those never reached their majority *in eighteenth-century Europe*. In America? We really can't tell, for although the clergy and government officials began to record vital statistics in the seventeenth century, births were unrecorded in the aggregate, and life-expectancy figures were not computed on a national level until this century. Reliable statistics exist for certain states, however. In Massachusetts, for example, the infant mortality rate during the second half of the nineteenth century fluctuated between 13 and 17 percent.

The great decline in infant mortality in Western society did not begin until the first few decades of the twentieth century, with improvements in the sterilization of food, the pasteurization of milk, the development of disinfectants, and modern medical advances. By 1915 neonatal deaths had declined to 4 percent in the United States, with only 9 percent of American infants failing to survive the first year. Succeeding decades brought even more spectacular gains in the industrialized countries of the world—and infant mortality rates continue to decline, as the accompanying table demonstrates.

INFANT MORTALITY RATES,*
SELECTED COUNTRIES:
1965, 1969, 1973

Country	1965	1969	1973
United States	24.7	20.9	17.7
Canada	23.6	19.3	15.5
Japan	18.3	14.1	11.3
France	18.1	16.4	†
Sweden	13.3	11.7	9.9

*Number of deaths of infants under 1 year of age per 1,000 live births.

† Not available.

Sources: Department of Economic and Social Affairs, United Nations, *U.N. Demographic Yearbook, 1974*; U.S. Department of Health, Education, and Welfare, National Center for Health Statistics, *Monthly Vital Statistics Report*, vol. 23, no. 11 and vol. 24, no. 11. Reprinted in U.S. Department of Commerce, *Social Indicators 1976* (Washington, D.C.: U.S. Government Printing Office, 1977), p. 213.

Unfortunately, throughout the less developed countries of the world the war against infant mortality has just begun. In

Upper Volta, for example, there are still 263 infant deaths per 1,000 live births.

AVERAGE LIFE-SPANS IN THE UNITED STATES

The 17 million Americans younger than 5 years could hardly care less about statistical tables, but as we grow older, life-expectancy figures do assume increasing significance, along with other vital statistics. No promises can be made, but the accompanying table shows how long a life-span you can expect if you've made it to a certain age.

LIFE-SPANS (YEARS) IN THE UNITED STATES, BY AGE, RACE, AND SEX

	Total		White		Nonwhite	
Age	Male	Female	Male	Female	Male	Female
0	67.6	75.3	68.4	76.1	61.9	70.1
5	69.2	76.7	69.9	77.3	64.1	72.1
10	69.3	76.8	70.0	77.4	64.3	72.3
15	69.5	76.9	70.5	77.5	64.5	72.4
20	69.9	77.1	70.7	77.7	64.9	72.6
30	70.9	77.5	71.4	78.0	66.7	73.3
40	71.8	78.0	72.2	78.5	68.7	74.4
50	73.4	79.1	73.6	79.5	71.5	76.4
60	76.2	80.9	76.2	81.1	75.6	79.3
70	80.4	83.6	80.4	83.7	80.7	83.2
80	86.4	87.9	86.3	87.9	87.9	89.4
85	89.8	90.8	89.7	90.7	91.3	92.3

Source: U.S. Department of Health, Education, and Welfare, *Vital Statistics of the U.S.*, vol. 2, *Mortality*, Part A, 1973, Table 5.1.

Here again, the accelerating advances of technology and science have had a tremendous effect on life expectancy. With each passing decade a few more years have been added to our lives, as can be seen in the accompanying table.

LIFE EXPECTANCY (YEARS) AT BIRTH IN THE UNITED STATES, BY RACE AND SEX: 1900–1974

Year	Total Both sexes	Male	Female	White Both sexes	Male	Female	Black and other races Both sexes	Male	Female
1900	47.3	46.3	48.3	47.6	46.6	48.7	33.0	32.5	33.5
1905	48.7	47.3	50.2	49.1	47.6	50.6	31.3	29.6	33.1
1910	50.0	48.4	51.8	50.3	48.6	52.0	35.6	33.8	37.5
1915	54.7	52.5	56.8	55.1	53.1	57.5	38.9	37.5	40.5
1920	54.1	53.6	54.6	54.9	54.4	55.6	45.3	45.5	45.2
1925	59.0	57.6	60.6	60.7	59.3	62.4	45.7	44.9	46.7
1930	59.7	58.1	61.6	61.4	59.7	63.5	48.1	47.3	49.2
1935	61.7	59.9	63.9	62.9	61.0	65.0	53.1	51.3	55.2
1940	62.9	60.8	65.2	64.2	62.1	66.6	53.1	51.5	54.9
1945	65.9	63.6	67.9	66.8	64.4	69.5	57.7	56.1	59.6
1950	68.2	65.6	71.1	69.1	66.5	72.2	60.8	59.1	62.9
1955	69.6	66.7	72.8	70.5	67.4	73.7	63.7	61.4	66.1
1960	69.7	66.6	73.1	70.6	67.4	74.1	63.6	61.1	66.3
1965	70.2	66.8	73.7	71.0	67.6	74.7	64.1	61.1	67.4
1970	70.9	67.1	74.8	71.7	68.0	75.6	65.3	61.3	69.4
1974	71.9	68.2	75.9	72.7	68.9	76.6	67.0	62.9	71.2

Source: U.S. Department of Health, Education, and Welfare, Public Health Service, National Center for Health Statistics, *Vital Statistics of the United States, 1973*, vol. 2; *Monthly Vital Statistics Report*, vol. 24, no. 11, supp. 1. Reprinted in U.S. Department of Commerce, *Social Indicators 1976* (Washington, D.C.: U.S. Government Printing Office, 1977), p. 190.

In the health-conscious 1970s changing habits are significantly affecting our projected longevity. According to Dr. Norman M. Kaplan of the University of Texas, reduced smoking and the growing tendency to avoid animal fats in our diets, along with better care and treatment of high blood pressure, have already added a couple of years to our life-spans.

Average Life-Span By Nationality

How does life expectancy in other parts of the globe compare with our own? It may come as a surprise that the United States does not even rank among the top 10 nations. Ahead of us are Sweden, the Netherlands, Iceland, Norway, Denmark, the Ryukyu Islands, Canada, France, Japan, and the United Kingdom.

There are astounding variations in life-spans around the world. A Swede, for example, can expect to live twice as long as a Nigerian and nearly three times as long as a male from Guinea, whose average life expectancy is only 26 years. These variations are attributable to many factors. The standard of living (itself determined by such crucial factors as the quality of medical care and nutrition), climate, cultural and social traditions (particularly the kinds of stresses found in each society) are only a few. The figures in the accompanying table, based on nationality at birth, were calculated in surveys after 1960, unless otherwise indicated. Where only one figure is given, the average life-spans of both males and females are combined.

LIFE-SPANS (YEARS) IN VARIOUS COUNTRIES: 1960

Country	Male	Female	Average: male & female
Angola			35
Argentina	63	69	
Australia	68	74	
Austria	67	74	
Belgium	68	74	
Brazil (1950)	45	51	
Burundi	35	38	
Cambodia (1959)	44	43	
Canada	69	75	
Central African Republic	33	36	
Chad	29	35	

Country	Male	Female	Average: male & female
Chile	54	60	
Czechoslovakia	67	74	
Denmark	70	75	
East Germany	69	74	
Ecuador	51	54	
England	69	75	
France	68	75	
Gabon	25	45	
Ghana			37
Greece	68	71	
Guinea	26	27	
Haiti			33
Hungary	67	72	
India	42	41	
Indonesia	48	48	
Ireland	68	72	
Israel	69	73	
Italy	67	72	
Japan	69	74	
Kenya			40–45
Liberia	36	39	
Mexico	58	60	
Morocco			47
Netherlands	71	77	
New Zealand	68	74	
Nigeria	37	37	
Northern Ireland	68	74	
Norway	71	76	
Pakistan	54	49	
Peru	53	56	
Philippines (1949)	49	53	
Poland	67	73	
Portugal	67	66	
Puerto Rico	67	72	

Country	Male	Female	Average: male & female
Rhodesia			
Africans			50
Europeans	67	74	
Scotland	67	73	
South Africa			
Whites	65	72	
Asiatics	58	60	
Blacks	50	54	
Spain	67	72	
Swaziland			44
Sweden	72	77	
Switzerland	69	74	
Taiwan	66	70	
Tanzania			40–41
Thailand	54	59	
Togo	31	38	
Turkey			54
United Arab Republic (Egypt)	52	54	
United States	68	75	
Upper Volta	32	31	
U.S.S.R.			70
Venezuela			66
West Germany	68	74	
Yugoslavia	62	66	

Source: Philip L. Altman and Dorothy S. Dittmer, eds. *Biology Data Book*, 2nd ed., vol. 1 (Bethesda: Federation of American Societies for Experimental Biology, 1972), pp. 224–226.

Life-Span and Occupation

In 1940, a 20-year-old American male could expect to spend the next 41.1 years of his life working—before retiring for the 5.7 nonworking years remaining to him. In 1970, his 20-year-old son could plan to work 5 months longer than his father—

and still have 8.8 years of retirement before the final reckoning.

It goes without saying that anyone concerned about wealth and happiness should choose his occupation carefully. But the job you take also influences your life-span. Professionals—doctors and lawyers—have the longest life-spans. Administrators, managers, and technicians follow; then farmers and agriculture workers; small businessmen, salesmen, and skilled workers; semiskilled workers; and finally unskilled laborers. Insurance companies and others who develop such statistics break down the categories even further. The life-span of a miner, for example, actually depends on the material he mines. Miners of metals and stone outlive petroleum workers, who, in turn, outlive coal miners.

For high-risk occupations, actuaries estimate the number of additional deaths per thousand that are likely per year: For astronauts, it is 30 extra deaths; race-car drivers, 25; aerial performers and professional prizefighters, 8. But for steeplejacks, it is only 3. So if you like speed, heights, or violence, be prepared for an early demise.

Another way to look at life-spans by profession is to consider the life-spans of the eminent in various fields of endeavor. Although not as wide ranging as the actuarial tables your insurance agent consults, the following list, which appeared in *Man*, by Richard J. Harrison and William Montagna, may help the budding genius or scholar to determine what path to follow.

Occupation	*Average life-span (years)*
Major symphony conductors	73.40
U.S. cabinet members	71.39
Inventors	70.96
Historians	70.60
Entomologists	70.39

Occupation	Average life-span (years)
American college university presidents	70.11
Geologists	69.79
Chemists	69.24
Educational theorists	69.06
Educators, all kinds	68.98
Economists and political scientists	68.68
Contributors to medicine and public hygiene	68.57
Botanists	68.36
Philosophers	68.22
Historical novelists	67.89
State governors (U.S.)	67.02
Authors of words to hymns	66.94
Mathematicians	66.62
Composers, grand opera	66.59
Composers, choral music	66.51
Composers, chamber music	66.26
Naval and military commanders (born 1666–1839)	66.14
Authors of political poetry	64.47
Painters in oil	64.22
British authors and poets	63.91

How do U.S. presidents rate? Except for the occupational hazard of assassination, being president is not as taxing to the health as it might seem. The job may have its pressures, but it apparently bestows benefits of longevity as well. Of the 36 U.S. presidents who are deceased, almost half—16 to be exact—lived to the age of 70 or older. Interestingly enough, two of the oldest presidents, Herbert Hoover and Harry Truman, remained in office through two of the most crisis-laden and stressful periods in U.S. history—the Depression and the years encompassing the end of World War II and the beginning of the Cold War. Here is a rundown of the 70+ club of presidents in order of historical tenure.

U.S. President	Term of service	Age at death
John Adams	1797–1801	90
Thomas Jefferson	1801–1809	83
James Madison	1809–1817	85
James Monroe	1817–1825	73
John Quincy Adams	1825–1829	80
Andrew Jackson	1829–1837	78
Martin Van Buren	1837–1841	79
John Tyler	1841–1845	71
Millard Fillmore	1850–1853	74
James Buchanan	1857–1861	77
Rutherford B. Hayes	1877–1881	70
Grover Cleveland	1885–1889 1893–1897	71
William Howard Taft	1909–1913	72
Herbert Hoover	1929–1933	90
Harry Truman	1945–1953	88
Dwight D. Eisenhower	1953–1961	78

European sovereigns, in contrast, have had a rougher time of it. Excessive intermarriage, court intrigue, frequent warfare, and the liberal use of poison and edicts of execution made the life of a ruling monarch a pretty hazardous affair in years past; hence, their meager average life-span of 49.14 years. Today, with monarchical power drastically diminished, they seem to be faring better in extending their longevity.

(For more on famous people and their life-spans, see the appendix at the end of this book.)

Life-Span and Marital Status

Strange as it seems, there is growing medical evidence that still other aspects of our lives influence our life-spans precisely because they have so much to do with whether we are happy or discontented. Emotional and physical well-being are more closely linked than many realize.

The connection is easiest to see in our relationships with other people. Loneliness and feelings of isolation are among the most potent forces that destroy our health and shorten our lives, a fact convincingly borne out by mortality figures in the United States. People who are single, widowed, or divorced are far more likely to die of a wide variety of illnesses than those who are married. Divorced women, for example, die of cervical cancer almost twice as often as married women. In many potentially fatal diseases, the strain of being divorced or widowed is one of the strongest predictors of death. Even being single is comparatively unhealthy. The death rate from heart disease is markedly higher for singles than for those who are married. From suicide and stroke, it is nearly twice as high; from cirrhosis of the liver, nearly three times.

In short, the old macho idea that people who need people are weaklings has nothing going for it. Medically speaking, everybody needs somebody just to survive.

Super-Codgers

Extravagant claims to longevity have been made by various people living in remote pockets of the earth. There have been people in the province of Hunza in Kashmir, in the Soviet Caucasus, and in the Andes of Ecuador who have claimed to be 125 years old or older. In most cases, skeptical outsiders have had no luck finding birth or baptismal records and so have had to accept these claims at face value. The one exception was the village of Vilcabamba in the Andes, where surviving baptismal records apparently certified these ages. Gerontologists, however, became suspicious when one man claimed to be 134 years old, a jump of 12 years in age from the last time he had been asked, 5 years before. Two University of Wisconsin researchers have now found that the confusion came about when records of grandparents and grandchildren with the same names were mistaken for one another. On a closer check of the records, the two skeptics concluded that no one alive to-day in the village could possibly be over 96 years old.

But while claims of reaching extremely advanced age have been exposed as false, it is nevertheless true that in certain remote mountain regions of the world unusual numbers of people—though plagued with minor health problems, such as rotten teeth—live to ripe and active old age. For example, according to the 1970 Soviet census, there were 2,500 centenarians residing in the Soviet republic of Azerbaidzhan, a figure representing 20 times the percentage of the population over 100 in the United States. The life histories of such people underscore the importance environment plays in the aging process. Dr. David Davies, a British gerontologist, spent several years in Ecuador among the "centenarians of the Andes" and isolated the following environmental factors he believes contribute to their longevity. They inhabit high-altitude regions near the equator, geographic areas having a dry climate and even temperature; they subsist on an austere, relatively unvaried diet low in calories, sugar, and animal fats, but with plenty of fruits and vegetables; they make elaborate use of medicinal herbs; they lead physically vigorous but unharried lives and are not subjected to the kinds of tensions and stresses experienced by individuals living in urban industrialized societies; there is little air pollution; the aged are not segregated and continue to fulfill important social functions. Furthermore, the aging process itself seems different. The elderly remain physically more youthful—and sexually active; but eventually they go into a swift decline as death approaches. A good deal more research needs to be done before any solid conclusions can be drawn.

For those still searching for the Fountain of Youth, but unwilling to move to Vilcabamba, the states with the greatest percentage of population over 65 are Kansas, Texas, Missouri, parts of Arizona and Oklahoma, and, of course, Florida; however, Mrs. Delina Filkins, the person with the world's longest authenticated life-span—113 years, 214 days—lived in New York.

Parts of the Body

PRESERVATION AFTER DEATH

When you're dead you're dead, the saying goes, and until recently there was little doubt about it, at least where the body was concerned. At one time it was thought that nails and hair continued to grow after death—a man who died clean-shaven would develop a stubbly beard, for example—but this is not true. What happens is that the skin recedes after death, causing the hair and nails to appear longer.

But even after the human body has stopped functioning and is considered dead, some of its parts can be preserved in states of suspended animation that may last a few hours or a few years—extending their life-spans separate from the body. Sophisticated cooling and freezing techniques have now ensured that tissues and organs that could save a life or help the blind see or even provide a baby to a childless couple need not die when their host organism expires, but can live on as parts ready to function for another human being. Because of our skill in preserving organs, over 30,000 people have received kidney transplants, and at least 1,400 children have been conceived with sperm that had been frozen and then thawed. As biologists perfect freezing techniques—usually by practice on animal parts—the out-of-body life-spans of disconnected organs and tissues are expected to get progressively longer. In fact, experiments with animals indicate that presently established shelf lives of human parts could probably be extended with little or no problem. To minimize medical risks, however, doctors prefer to err on the conservative side in setting these limits. Here are the standard life-spans they use.

Blood. The Red Cross stores refrigerated blood for 3 weeks; after that time the blood is turned over to laboratories to be used in research.

Corneas. Maximum storage life is usually 3 to 4 days in a tissue culture, although there have been instances where corneas frozen for longer periods were safely transplanted. Maximum frozen storage life ranges from 6 months to 1 year. Some corneas are stored in glycerin for as long as 1 year and are used, not as cornea replacements, but as natural "bandages" for damaged eyes awaiting new, fresh corneas.

Kidneys. There are two techniques for storing a fresh human kidney. Immediately after being removed from the body, the kidney can be placed in cold storage—a cold electrolyte (potassium or chloride) solution—and then be preserved for up to 18 hours. In the pulsatile perfusion technique, a cold, oxygenated fluid is pumped through the kidney, quickly cooling and preserving it for as long as 72 hours.

Liver. An electrolyte solution pumped through a human liver packed in ice will maintain the organ in transplantable condition for 2 to 8 hours as a general rule, but at least one liver has been stored for 16 hours in this way with no complications in transplantation.

Heart. A live human heart can be kept in perfect condition for as long as 4 hours after removal if an electrolyte solution is pumped through its blood vessels. (The real storage limit may be much higher: Animal hearts have been successfully transplanted after 24 hours.)

Bone Marrow. Stored at an icy –50° C, marrow can be kept fresh for 6 months to 1 year.

Sperm. Healthy, normal babies have been carried by mothers artificially impregnated with sperm that had been stored frozen for as long as 13 years. (Again, the real storage limit is probably much higher: Bull sperm frozen as long as 25 years has produced healthy, normal calves.)

The Brain and Beyond. Looking toward the future, one Japanese doctor at Kobe Medical College managed to revive a cat's brain that had been frozen for 7 months.

IN-BODY LIFE-SPANS

Except for organs and tissues kept alive by extraordinary medical means, the various elements of the human body have life-spans that are the same as, or shorter than, the life-span of the body itself. And with respect to certain body parts we can compute individual life-spans.

Body Cells. The life-spans of the various types of cells that together make up a human being vary considerably. Among those with the shortest life-spans are blood platelets (4 days), white blood cells (9 days), and skin cells (20 days). Red blood cells live for about 120 days. In contrast, muscle and nerve cells may live as long as their host organism.

The male sex cells, sperm or spermatozoa, which are continually being produced in the testicles, mature as they travel through the epididymis (a coiled tube several yards long leading from the testicles to the vas deferens), a journey lasting about 3 weeks. After ejaculation, a sperm can remain viable for 2 to 5 days in the female reproductive tract.

In contrast, the female sex cells, the ova or eggs, are all present—up to 700,000 of them—at birth. During ovulation, usually one egg is released from one of the ovaries. This egg remains viable for only about 12 hours unless it is impregnated. Sometimes two eggs are released and, if impregnated, can produce fraternal (dizygotic) twins. When more than two eggs are released, higher multiple births can result.

The various kinds of cells in the body are not equally vulnerable to the aging process. Essentially, cells that continue to undergo mitosis—that is, reproduction by means of cell division—retain their youthfulness. They do not normally show signs of degeneration or impaired functioning. In contrast, postmitotics—such as heart muscle cells or neurons—are incapable of further division; they cannot rejuvenate or replace themselves. As they age, they accumulate pigment and other substances, such as calcium or iron, that cause them eventually to deteriorate; or they may fragment and undergo other changes in their nuclei that result in impaired functioning or death.

Hair. External human hair—that part of the hair that extends beyond the surface of the skin—is already dead. Biologically speaking, it has exhausted its life-span, although the body will not shed it for quite a while. Human hair grows in cycles. An individual hair grows for a period of time (up to 6 years) and then goes into a quiescent stage for 1 to 3 months. At the end of this dormant period a new hair begins to grow—at a rate of about 0.33 millimeter per day—pushing the old one out. We lose as many as 300 hairs a day in this fashion. In most mammals this process is seasonal, most of the hairs going through their respective growth and quiescent periods at the same time. In human beings, however, the process is continual—some hairs growing, while others are quiescent.

Unless, of course, you are in the process of growing bald. Advancing age—with its attendant hormonal changes—is the common cause of male pattern baldness, the medical term used to describe normal loss of hair. Although the hair follicle itself remains alive, changes in the level of androgens cause it to cease to grow hair. No amount of scalp massage or application of vitamins or exotic ointments will reverse this process; scalp punch autograft (hair transplants) will only redistribute the remaining hair-producing follicles to give the appearance of a full head of hair. Whether or not you will grow bald is a genetic matter. To determine what the future has in store, look at your maternal grandfather, since the responsible gene, on the sex-linked chromosome, skips a generation.

There is another type of baldness that is neither hereditary nor irreversible. Excessive hair loss—manifested either in localized bald spots or in overall thinning—is caused by extreme deficiencies in nutrition (particularly iron), excessive amounts of vitamin A, certain medicines that upset hormonal balances, illness, physical trauma, or childbirth. Excessive stress, depression, and other mental illnesses can also precipitate hair loss. Such abnormal hair loss can be reversed by eliminating the conditions that initially triggered the malfunction. If poor diet is the problem, hair loss will slow when nutrition is improved because hair follicles are among the most rapidly metabolizing structures of the body. A few months on a balanced diet and normal hair growth will be restored. Women whose hair becomes thin on account of childbirth or menopause benefit from estrogen therapy, although such treatment may involve health risks. A safer bet might be to invest in a wig or choose a hairstyle that gives the appearance of fullness, until the body returns to normal hormone levels.

Nails. Finger- and toenails begin to grow before birth and continue throughout life. Healthy nails grow about 0.8 inch each year. The standard unit of nail growth, however, is called the nail-second. This figure, the amount a nail grows in a single second, is .0000039 inch.

Teeth. Most of us begin to develop our deciduous, or baby, teeth at about 6 to 9 months of age. (Louis XIV, on the other hand, was *born* with two teeth.) They begin to fall out at 5 to 7 years of age, the upper ones first. With the exception of the permanent third molars, or wisdom teeth, our adult or permanent teeth have usually all arrived by the time we are 12 to 15 years of age.

Human Functions

Strictly speaking, the length of time a human function lasts is a duration, not a life-span, but the following activities are so much a part of our lives, they should be included.

RESPIRATION

At rest, we breathe about 16 times a minute; hence the average "life-span" of a breath is 3.75 seconds. The rate of breathing is controlled by the brain's respiratory center, which responds to changes in the level of hydrogen ions and carbon dioxide in the blood, as well as to other factors such as stress, temperature change, and motor activities.

Each time we breathe, we inhale a mixture of gases that is approximately 76 percent nitrogen, 23 percent oxygen, and 1 percent argon. (There are other elements in the atmosphere, but only in minute quantities.) A few seconds later, we exhale nitrogen, carbon dioxide, and argon. The nitrogen and carbon dioxide eventually form other compounds; the argon remains just as it was. The celebrated astronomer Harlow Shapley has noted that each breath contains billions upon billions of atoms of argon (actually 3×10^{19}); when exhaled, these atoms begin to spread through the atmosphere. Within a year, the atoms are evenly distributed throughout the atmosphere. We are sure, therefore, to be breathing the same ones again. As Shapley points out, this has curious implications. With every breath we take we are inhaling some of the same argon atoms breathed by Socrates, Caesar, and Abraham Lincoln.

SLEEP

Sleep is not a fixed commodity in your body's cycle. As you age, how long you sleep will change. The accompanying table gives a general idea of how long a night's sleep commonly lasts.

Age	Sleep time (hours)
1–15 days	16–22
6–23 months	13
3–9 years	11
10–13 years	10
14–18 years	9
19–30 years	8
31–45 years	7.5
46–50 years	6
50 + years	5.5

DREAMS

Until recently it was thought that dreams last about 1 hour and that we dream four or five dreams per night. Current theories, however, suggest that there are flash or instantaneous dreams, on the one hand, and dreams that last throughout the night, on the other. Most dreams last as long as they seemed to last upon awakening—the duration of a dream corresponds to the amount of time it would actually take to imagine the activity.

REPRODUCTION

Female Reproductive Life. Human females can produce children from the onset of puberty (11 to 15 years of age) until menopause (45 to 50 years). It is extremely rare for a woman over 50 to give birth. (Albania has the highest rate for such births—about 5,500 per million.) There have been reports of children born to women in their sixties and seventies, but the oldest woman in an authenticated instance is Mrs. Ruth Kistler of Glendale, California, who gave birth to a baby girl when she was 57 years, 129 days old.

Male Reproductive Life. Human males can sire children from puberty (12 to 16 years) until late in life. Men undergo no true menopause, although their testicular activity declines slowly over the years. While it is extremely difficult to authenticate paternity, there are documented instances of men in their nineties siring children.

Your Own Life-Span*

Only seers and tea-leaf readers can claim to know for sure how long you'll live, but you can make a reasonable stab at estab-

*This section was adapted from "Will You Live to be 100?" by Judith Bentley, *Family Health*, January 1975.

lishing your own life-span with this enlightened exercise in addition. Start by looking up your statistical life-span on the age/sex/race table on page 11. Taking that number, apply the following factors.

HEREDITY

1. If two of your grandparents lived past 80, add 2 years to your life expectancy. If all four made it to 80 or beyond, add 5.

2. If anyone in your immediate family died of heart attack or stroke before age 50, subtract 4.

3. For a parent, brother, or sister who has cancer, a heart condition, or diabetes, subtract 3.

LIFE-STYLE

1. If you live in a large (2 million plus) city, subtract 1; if you live in a small (10,000 or under) town, add 1.

2. If your work keeps you sitting behind a desk, subtract 2; if it involves physical activity, add 2.

3. If you are living with someone, add 4; if not, subtract 1 for each decade alone since 25.

4. If you use seat belts and stay within speed limits when driving, add one; if you got a speeding ticket this year, subtract one. (Auto accidents are among the top five killers.)

5. If you make over $50,000, subtract 2. (High-income jobs yield short life expectancy.)

6. If you are a college graduate, add 1; if you received any post-graduate degree, add 2.

7. If you're 60 or older and still working, add 3.

HEALTH HABITS

1. For smokers: If you consume two packs a day, subtract 8; one to two packs, subtract 6; one pack or less, subtract 3.

2. For every ten pounds you are overweight, subtract 1.

3. For exercising three times weekly or more, add 3.

4. Checkups: for men over 40, add 2 if you have annual check-ups; for all women who have at least one checkup annually, add 2.

5. If you sleep 9 hours or more, subtract 4.

ATTITUDES

1. For the intense, aggressive, easily angered person, subtract 3; for the easygoing, passive person, add 3.

2. For those unhappy with life, subtract 2; for those who are happy, add 2.

The number you get gives a reasonable estimate of your life expectancy based on what is known about various factors and how they work for or against us in everyday life.

2 / THE ANIMAL KINGDOM

HUMAN beings, with their life-spans of 70+ years, are relatively long-lived members of the animal kingdom. Elephants, alligators, giant salamanders, hippopotamuses, rhinoceroses, gorillas, whales, and a scattering of birds and fishes have life-spans that approach man's. But the only animals that live longer than man are the tortoises, the turtles, the European freshwater clam, certain bacteria, and the so-called immortals—simple, multicelled creatures like the sea anemones (flower-animals) and the sponges.

The life-span of the immortals is indeterminate—theoretically, they can go on forever. One colony of sea anemones, for example, was kept in a laboratory for nearly 100 years, always reproducing and never displaying degenerative changes. Its ultimate demise is thought to have resulted from changes in the laboratory environment rather than from biological aging.

The record for animal life belongs to the black Seychelles tortoise, a species now extinct. One of these tortoises was captured on the Seychelles Islands by the French explorer, Marion de Fresne, in 1766 and was then transported to Port Louis, on the island of Mauritius, where it was placed in an artillery barracks. There it lived until 1918, when, seemingly in excellent health, it fell through a gun emplacement and died—152 years after its capture. At capture, the tortoise was already an adult, making its final actual age at least 170, perhaps more. There have been reports of even older tortoises, including a supposedly 200-year-old tortoise brought from Madagascar to

Tonga by Captain Cook, but the reports are not thought to be reliable.

One theory holds that the larger the animal, the longer its life-span. For the most part, this proves true. Tortoises, for example, can weigh more than 500 pounds. (The notable exception is the Carolina box turtle; although it is only about 6 inches long, it outlives all but the oldest tortoises. One of these little turtles was tagged, set free, and then captured again— 129 years later.)

Sex also plays a role in longevity. The animal kingdom is, in many instances, a male world; but in terms of life-spans, females have the upper hand. In almost every species—from fish to poultry, from housefly to man—females outlive males, often substantially so. And it is not simply a case of males leading more stressful lives; there is actually a higher rate of male mortality in the uterus and in infancy, as well as in adult life.

Little is known about why animals live as long as they do. Those scientists who believe that each cell of an animal contains a biological clock that determines the time of death have calculated the number of times particular cells are able to divide to see whether such figures correspond to the animal's life-span. The tortoise has cells that divide 90 to 125 times; the chicken, a creature of medium life-span, has cells that divide 15 to 35 times; the mouse, a short-lived animal, has cells that divide 14 to 28 times. Our own cells divide 40 to 60 times. The results support, but do not prove, the theory.

The life-spans of some animals, among them rats, have been prolonged in laboratory experiments by putting the animals on near-starvation diets early in life. More recently, biologists have found that placing middle-aged mice in a low-temperature environment has tripled the mice's life-spans. Other experiments suggest that keeping an animal from mating also prolongs life. (However, wild animals generally succumb not long after they lose their reproductive capacities.)

Despite the enormous amount of biological and medical research involving animals, determining their life-spans is more an exercise in creative deduction than in investigation. Indeed, for many animals, our knowledge is little better than it was in the 1600s, when Sir Francis Bacon noted: "The knowledge ... is slender, the observation negligent, and tradition fabulous; in household beasts the idle life corrupts; in wild, the violence of the climate cuts them off."

The problem, of course, is that most animals have two life-spans: one in their natural habitats and one in captivity. The impossibility of monitoring wildlife closely (if it were monitored, it would not really be wild) has forced zoologists to confine their determinations of life-spans in the wild to educated guesses, estimates based on a small sample of tagged animals, and such physical evidence as size, number of teeth, and growth rings (in shells, bones, the wax earplugs of whales, and the scales of fishes). The accuracy of these indicators is open to question; moreover, few animals actually live to an old age in the wild, and we are forced to predict life-spans based upon "the oldest known example" or on averages of widely divergent ages.

Animals in captivity, of course, have easily verifiable life-spans. But the conditions of captivity are such that we can expect little correlation between the length of time an animal will live safely and securely in a zoo and the length of time it would survive in the wild. The passerine, or perching, birds, for example, can live up to 20 years in captivity; they seldom reach their second birthdays in the wild.

The science of keeping animals alive and happy in a zoo is a complicated one, demanding the wide-ranging knowledge of a cosmopolitan naturalist, the medical skills of a veterinarian, and the instincts and experience of a social director on a cruise ship. Every animal has its own physical and psychological needs that affect how well it does away from its home turf. Some animals are simply more adaptable than others; as a result, they enjoy longer life-spans when confined in zoos.

As zookeepers have discovered over the years, keeping an animal alive in captivity is a tricky business, full of surprises.

For example, a spiny anteater housed at the Philadelphia Zoo was, to all appearances, a miserable creature—a nocturnal, extremely shy animal that spent the better part of its zoo life huddled in a 3-foot-square wooden box away from the prying eyes of the zoo's visitors. This supposedly distraught animal lived to be well over 50 years old, making it one of the oldest on record in the animal kingdom. On the other hand, there are spider monkeys that apparently adjust extremely well to captivity and thrive on the attention they get from visitors. As long as they have room to scamper around and other monkeys to play with, they appear happy. Psychologically, they couldn't be more ideal animals for captivity. Unfortunately, they are not very hardy physically and are prone to disease and infections. For that reason, they usually live only 4 to 5 years in captivity, although there have been rare cases of spider monkeys living as long as 18 to 20 years.

These variations occur even within the same family of animals. The jaguar is notorious for making a poor psychological adjustment to zoo life, in which he is inclined to be moody, grumpy, and unpredictable. In contrast, the cheetah thrives on human attention and thus is very easy to tame. In spite of these differences, jaguars routinely live 13 to 25 years, while the more delicate cheetahs seldom make it past their sixth birthdays.

Some animals, it seems, do too well in captivity and live long past their prime. Lions are the best example of this; there are records of lions living as long as 29 years in zoos. Lions tend to get senile before they get sick—usually somewhere between their tenth and fifteenth years. Zookeepers are often forced to put the senile cats to sleep.

The impact of zoo life on life-spans can therefore be quite complex. Recent trends in designing zoos to recreate natural habitats undoubtedly ease the animals' psychological adaptation, and special diets and care improve their health. But ultimately the measure of an animal's life-span is locked within the animal itself.

Mammals

The longest-lived mammalian species is man. Of the others, only the elephant, horse, hippopotamus, and whale have been able to approach 50 years with any regularity. Most medium-sized and large mammals seldom breach the barrier of 30; the larger rodents and smaller carnivores hardly ever make it past 10 (mice and shrews have trouble cracking 5). The life-spans of mammals generally relate to size: The larger the animal, the longer it lives. Man, again, is the major exception, but man has the obvious advantage of brain power. There isn't an elephant around that can discover penicillin, invent wheelchairs and pacemakers, or establish health spas to prolong the lives of its fellows.

In the mammalian world, the age records of horses, men, and domestic pets are considered the most reliable, since these are the only species we have been able to study in large enough samples to ensure accurate estimates. Because of the problems of studying life in the wild and the unnatural effects of captivity, all other life-spans are educated guesses.

As a general rule, mammalian females live longer than mammalian males. But longer life doesn't necessarily mean fuller; in a recent study of mice, it was found that virgin females lived longer than spayed females, who, in turn, lived longer than sexually active females.

AARDVARKS

Also called ant bears and earth pigs, these long-eared, long-nosed African anteaters do not adjust well to captivity, but examples have lived for up to 10 years in zoos.

ANTEATERS

The term *anteater* refers to a variety of mammals that, of course, feed on ants. The South American giant anteater has survived 14 years in captivity; the spiny anteater of Australia, New Guinea, and Tasmania, 50 years; the long-nosed anteater of New Guinea, 30 years.

ANTELOPES

There are more than 90 species of antelopes, ranging from the royal antelope, only 12 inches high at the shoulder, to the giant eland, which stands 6 feet at the shoulder and can weigh 1 ton. There are few species of antelopes whose life-spans are known. Eland have life expectancies of 15 to 20 years in the wild. Gazelles—the dozen or so species of small and medium-sized antelopes found in Africa, India, Mongolia, and the Middle East—have life-spans of 10 to 12 years.

ARMADILLOS

Armadillos are a singular group of armored, so-called toothless mammals. Without their protective plates and the ability to quickly dig holes in which to hide, they would have disappeared eons ago. They have been known to survive more than 15 years in captivity; a hairy armadillo lived in a Zurich zoo for over 17 years.

ASSES

Asses have probably lived to 40. The report of an 86-year-old ass in the *London Times* on November 29, 1937, can hardly be taken seriously.

AYE-AYES

Found only in Madagascar, this long-eared, monkeylike species of primate, about the size of a house cat, can live up to 23 years in captivity.

BABOONS

Although there are instances of baboons living more than 40 years in captivity, 20 years is thought to be their maximum life-span in the wild. Their adaptability and well-ordered communal life, in which "knowledge" is handed down to younger generations, have provided sufficient protection to these smaller members of the African wildlife scene for them to have survived as a species for 35 million years.

BADGERS

The American badger, found from Canada to central Mexico, has lived up to 13 years in captivity. The Eurasian or common badger, found in Europe and Asia, can live up to 15 years.

BATS

The only flying mammals, bats can have life-spans of 20 years or longer. The common vampire bat, for example, is known to live up to 12 years in the wild. A fruit bat owned by the London Zoo lived for a record 17 years, and fruit bats have been known to live over 20 years in the wilds of Madagascar, South East Asia, and the Philippines. The longest-lived bat on record was a small brown example that was banded in Massachusetts in 1937 and found again in Vermont in 1960, making it at least 24 years old. A bat lives so much longer than its small-rodent relatives possibly because it is able to reduce its temperature and metabolic rate while resting in the daytime, and some species hibernate through cold winters. Perhaps to compensate for their longevity, bats are slow breeders and have only one or two offspring each year.

BEARS

The life-span of a bear—black, brown, polar, and grizzly alike—is 15 to 34 years in the wild. There have been instances of brown bears living up to 47 years in captivity; polar bears, 41 years; and grizzlies, 31 years. The celebrated Smokey the Bear, symbol of the National Park Service, died on November 9, 1976, at age 26.

BEAVERS

These large rodents—beavers can weigh up to 50 pounds—live 15 years in the wild and up to 30 years in captivity.

BOBCATS

These small North American lynxes live up to 20 years in the wild, unless farmers kill them to protect livestock and poultry. Bobcats are very comfortable in captivity: They have been known to survive in zoos for over 30 years.

BUFFALOES

The Asiatic buffalo (also known as the Indian or water buffalo) and the large African buffalo (prized by big-game hunters) have life-spans of 10 to 20 years. Examples of the American bison (often erroneously called a buffalo) are now kept in protected herds, where they number about 30,000 and enjoy life-spans of 18 to 22 years.

CAMELS

Both the two-humped Bactrian camel and the one-humped dromedary may live more than 45 years, although a 25-year-old camel is considered old.

CAPYBARAS

The world's largest rodent (the capybara can weigh up to 125 pounds) resembles a giant guinea pig. It has a life-span of 8 to 10 years in the wild, although some have lived over 12 years in captivity.

CATS

The longest-lived of the small domestic animals, cats have life-spans of 13 to 17 years. Some live much longer—there are authenticated cases of cats over 30. According to a study done at the University of Pennsylvania School of Veterinary Medicine, two factors affect a cat's life-span: whether it has been sterilized and whether it is a purebred or of mixed breed. In one study of 629 cats, the researchers found that the life-span, particularly of males, benefited from neutering. Males that were castrated lived an average of 10.8 years, while those left untouched managed only 8.6 years. Interestingly, being spayed made little difference in how long females lived. A gonadectomy, veterinarians say, may help raise a cat's resistance to infection (perhaps accounting for the longevity effect on males). Unfortunately, it also seems to increase the animal's susceptibility to cancer, the leading cause of death among the older females (and perhaps accounting for its lack of effect on females). In general, mixed-breed cats live about 3 months to 1 year longer than the less hardy purebreds, who seem to be more susceptible to infection and disease.

CATTLE

Domesticated cattle have potential life-spans of slightly more than 20 years. It is a rare animal that even approaches this limit, however. Cattle and steers raised for beef are usually slaughtered around their second birthdays. Dairy cattle are kept only until the quantity and butterfat content of their milk begin to drop, then are culled from the herd. This happens at age 8 or 9.

CHEETAHS

With its long slim legs, the cheetah looks more like a dog with a cat's head than a member of the cat family. The swiftest four-footed animal alive, it can run up to 65 miles per hour in short bursts. Cheetahs have been known to live for up to 14 years in captivity, but their chances for survival after capture are usually poor, particularly if they have been caught young. In a survey made in 1967, the average zoo life of the cheetah was a mere 3.5 years.

CHIMPANZEES

Chimps are thought to have life-spans of 35 to 40 years. The females follow the rule of tending to outlive the males, but the longest authenticated life for a chimp belonged to a male named Heine of the Lincoln Park Zoo in Chicago. Heine was 50 years old when he died.

CHIPMUNKS

Found in various forms throughout the world, chipmunks live about 5 years in the wild.

COYOTES

Despite their reputation as killers of livestock and the resulting persecution by man, coyotes still manage to survive about 10 years in the wild. As a species they are extremely hardy, and some zoologists predict the coyote will inhabit this planet long after man has gone.

CUSCUSES

These sluggish, woolly-haired, slothlike mammals inhabit the trees of New Guinea and Australia. In zoos, they have been known to live over 11 years.

DEER

The 41 species of deer have life-spans ranging from 10 to 20 years. Caribou, the only species in which both sexes have antlers, live up to 15 years in the wild, for example.

DOGS

Few domestic dogs reach the age of 20. In general, small dogs live longer than large dogs—a reversal of the usual trend in mammals—although the record for longevity belongs to a fairly large dog named Adjutant, a black Labrador retriever. Adjutant lived to be 27, growing old in the care of his lifetime owner, James Hawkes, an English gamekeeper.

The old saw of correlating a dog's age to human life-span by multiplying his actual years by seven has been shown to be false. A recent study of beagles revealed that, while about one-third died by 10 years of age, signs of true senility were present from the age of 5 on.

The following is a short list of assumed or observed life-spans of various breeds of dogs. To determine the statistical life-span of your dog, assuming it is not one of these breeds, simply split the difference between the two breeds immediately larger and smaller than your pet. Your dog's potential span, whether it is male or female, spayed, neutered, or normal, will be right there:

Life-span (years)

Pekinese	20
Dachshund	19
Fox terrier	16
Mastiff	14
Saint Bernard	14

ELEPHANTS

The life-span of both the African and the Asiatic elephant is 60 to 65 years. The oldest elephant in captivity was a female Indian or Asiatic elephant named Jessie, who lived in the Taronga Park Zoo in Sydney, Australia, for 57 years and was thought to have been 12 to 20 years old when she arrived, making her 69 to 77 years old when she died.

FOXES

Most foxes in the United States are trapped and seldom survive more than 3 or 4 years. In captivity, however, foxes can live 12 to 14 years.

GERBILS

The 100 or so species of this desert rodent are similar to rats in both size and appearance. They are popular pets and live up to 5 years.

GIBBONS

Because of the surprisingly human look of the gibbon embryo, the smallest of the apes was once thought to be man's closest primate relative. Gibbons now are recognized as a separate family, different from both man and the great apes. Gibbons can live more than 30 years in captivity.

GIRAFFES

The tallest animals on four legs, giraffes have life-spans of 15 to 20 years, but they have survived for up to 28 years in captivity. Their long necks not only enable them to reach edible vegetation but also to keep their throats away from possible predators, and a swift kick from a giraffe can decapitate a lion.

GOATS

Goats knock about for 15 to 20 years, despite diets that include anything from woolen clothing to tin cans.

GOPHERS

These rodents live from 2 to 5 years.

GORILLAS

Not that much is known of gorillas in their natural habitats, the forests of central Africa, but their life-spans are estimated to be 25 to 30 years. The oldest gorilla now in captivity is Massa, who, as of April 1, 1979, was 48 years old. Massa has lived at the Philadelphia Zoo for the last 43 years.

GUINEA PIGS

Guinea pigs are not true pigs; nor are they from Guinea. They are, in fact, rodents and were kept as domestic animals in Peru long before the Spaniards conquered the Incas. A guinea pig that manages to escape scientists and researchers can live about 7 years in captivity.

HAMSTERS

Natives of Europe, parts of Asia, and the Middle East, hamsters have become popular in the United States as laboratory animals and pets. The life-span of a hamster is about 3 years.

HEDGEHOGS

Hedgehogs are either spiny or hairy. They have an extraordinary ability to roll themselves up into tight balls when frightened, making themselves invulnerable to almost all predators. Nonetheless, they live for only 6 years.

HIPPOPOTAMUSES

In a study conducted in Rwenzori National Park in Africa, the life-span of the hippopotamus was found to be about 43 years. The record for hippopotamuses in captivity is held by Peter the Great, of the Bronx Zoo, who lived for 49 years, 6 months. A hippo's fondness for mud is no joke: Lacking any external body mechanism for resisting heat, hippopotamuses submerge in mud baths for 5 minutes at a stretch to lower their body temperature.

HOGS

The domestic hog has a life-span of about 20 years. He is outlived by the wild hog, which can live up to 27 years.

HORSES

Generally, horses have life-spans of 25 to 30 years, although Arabians consistently live up to 32 years of age. There are a number of reports of horses over 40; The *Guinness Book of World Records* lists Old Billy, an English horse who supposedly reached the age of 62.

HYENAS

Hyenas may look like dogs, but they are not in the canine family. Though the hyena may appear clumsy, it can outrun a cheetah after the first 0.5 mile. Nor are hyenas the scavengers legend makes them out to be; they are excellent primary hunters. They live about 25 years in the wild now that there is no longer a bounty on their heads.

JAGUARS

The jaguar, somewhat larger than the leopard and with a shorter tail, lives up to 22 years.

KANGAROOS

The oldest authenticated age for a kangaroo is 20 years—for a grizzled-gray tree kangaroo from the National Zoological Park in Washington, D.C.

KOALAS

Koalas are not really the tiny animals depicted in the Qantas commercials. The largest of the so-called phalangers (fingered marsupials), they can weigh up to 35 pounds—on a diet consisting mostly of eucalyptus leaves—and live as long as 20 years.

LEMMINGS

The life expectancy of a lemming is only 1 to 1.5 years; in very rare cases, 2 years.

Lemmings normally make short annual spring migrations in search of new sources of food and shelter. But after a 3- to 4-year period of high birth rate and low mortality, they begin a massive migration. Exactly why this happens is not quite clear; it appears to be triggered by lessening food supplies, overcrowding, and the resulting increased competition for the necessities of life. Whatever the reason, lemmings then begin to move in huge numbers, traveling night and day. In areas like Scandinavia, where the mountainous homes of the lemmings are close to the sea, the animals' mad dash can take them right to the water, where those who have survived the trek plunge into the surf and swim until exhaustion sends them under.

LEMURS

Common in U.S. zoos, the lemur is a kind of primitive monkey, which can live more than 20 years in captivity.

LLAMAS

Smaller than its camel cousin (and lacking a hump), this hardy mountain animal lives only about 20 years.

LYNXES

Lynxes once roamed throughout Europe but have become scarce as a result of intensive hunting by greedy trappers. In secure zoos they live for up to 10 years.

MARMOTS

Marmots are the fat little furry animals seen in many national parks in the western United States. They are, in fact, woodchucks and have life-spans of up to 15 years.

MICE

Common house mice have lived up to 6 years, but the average life-span is less than 5.

MINKS

Minks can live for up to 10 years in captivity, if they have been designated as breeders. If not, they live only from May to December, when they are killed and immediately skinned.

MONGOOSES

Mongooses are highly agile animals that are kept as rat and snake killers in many countries. They can live for about 15 years in captivity.

MONKEYS

The life-spans of monkeys differ by species. The marmoset, the smallest of the monkeys, lives about 10 years in the wild and up to 16 years in captivity. The rhesus monkey has a life-span of about 20 years; the squirrel monkey, about 15 to 20 years; the macaques, up to 30 years. The white-faced capuchin monkey (the organ grinder's monkey) is particularly long-lived; at the Mesker Park Zoo in Evansville, Indiana, a white-faced capuchin named Jerry lived to be 47 years old.

MOOSE

This largest animal in the deer family is also blessed with a relatively long life-span—about 20 years—to say nothing of its postmortem "life" hanging on a hunter's wall.

MOUNTAIN LIONS

The mountain lion, or puma, has few natural enemies besides man. Unfortunately, man is such an efficient enemy that the puma has almost disappeared from the earth. In captivity, pumas can live as long as 19 years.

OPOSSUMS

These nocturnal marsupials, which play dead when cornered, don't have long before they deal with the real thing; an opossum's life-span seldom exceeds 2 years.

ORANGUTANS

In captivity orangutans have lived over 30 years, but it is thought that their life-spans in the wild may be as long as 40 years.

OTTERS

Playful members of the weasel family, otters have a tough time of it; relentless hunting of several species has led to their near extinction. If the trappers don't get to them, they can hope for up to 6 years.

PECCARIES

These small wild pigs—the only species of pig native to the Americas—are ferocious fighters when threatened; they live about 20 years.

PIGS

The farm variety of pig lives anywhere from 10 to 20 years, even though pigs are susceptible to a greater number of diseases than any other domestic animal. Many of these diseases, such as trichinosis and brucellosis, are transmissible to man.

PLATYPUSES

Also called duckbills or duck-billed platypuses, these primitive egg-laying mammals live up to 14 years.

PORCUPINES

These slow-moving rodents take it easy. They have 10 to 15 years in the wild; up to 20 years in captivity. Porcupines cannot actually "shoot" their quills. The quills are loosely attached and pull out easily, remaining embedded in any predator that comes in contact with them.

PORPOISES

Although they are also called dolphins, porpoises belong to the mammalian order of whales and are not related to the dolphin fish. They are intelligent, delightful creatures, empathetic in both fact and myth, that have been known to save individuals from drowning and even aid ships in danger of being beached. They have been rewarded for their good works with relatively long life-spans: Depending on the species, porpoises live 25 to 50 years.

PRAIRIE DOGS

Although they may bark like dogs, prairie dogs are members of the squirrel family. They need never fear death by dehydration: Equipped with a special chemical process that transforms solid food into water, they do not drink. Nonetheless, they live only 10 years.

RABBITS

Rabbits have an average life-span of less than 2 years in the wild and a maximum life-span of about 5 years. The same rabbits, in captivity, can live at least 10 years. Domesticated strains have longer life-spans, however, and there are instances of females surviving to the remarkable age of 18.

RACCOONS

Omnivorous creatures, raccoons find it very easy to adapt to civilization, enjoying its fruits—whether in the garden or the trash can—for about 10 years.

RATS

Rats live up to 6 years, but they make the most of it, conceptionally speaking: One female can produce 800 offspring.

REINDEER

Reindeer live about 15 years. Migratory and gregarious by nature, they travel in herds of up to 200,000 animals and are capable of covering 40 miles a day.

RHINOCEROSES

These thick-skinned mammals may seem stupid, but they live up to 50 years, despite the fact that they adjust poorly to civilization and are hunted for their horns, which, in China, have for centuries been considered a potent aphrodisiac.

SEALS

The average life-span of the fur seal is 7 years in the wild, although seals have been kept in captivity up to 26 years. Harbor seals can live more than 30 years, and true, or earless, seals such as the gray seal can live more than 40 years (one, shot in 1969, was believed to be at least 46).

SEA LIONS

Sea lions differ from seals in their size, their roaring bark, and their lack of underfur. They can live up to 23 years in captivity, depending upon the species.

SHEEP

The life-spans of domestic sheep are about 7 years. Wild sheep such as the bighorn, however, live longer—up to 20 years for the rams and as long as 23 years for the ewes. The strongest and largest rams seem to die first, supposedly because they breed with the largest number of ewes—a dangerous sport that involves a great deal of fighting with other rams, leaving little time for eating.

SHREWS

These primitive mammals live only 12 to 18 months in the wild, mainly because they have so many enemies. While they are around, however, they kill and consume twice their own weight each day.

SKUNKS

Skunks are members of the weasel family and live up to 10 years in captivity.

SLOTHS

The two-toed sloth can live—very slowly—for 11 years in captivity.

SQUIRRELS

These backyard animals can live up to 15 years in captivity.

TASMANIAN DEVILS

About the size of a large cat—but with a 12-inch tail and an evil expression—this burrowing marsupial has a life-span of about 8 years.

WALLAROOS

A kind of kangaroo, the wallaroo has a life-span of 15 to 20 years.

WALRUSES

Walruses live up to 40 years in the wild.

WHALES

The average life-span of a whale is about 50 years. Sperm whales in the northern Pacific, however, are thought by some sources to live upwards of 70 years. Then there is the case of Old Tom, a killer whale with peculiar markings that was sighted along the Australian coast over a period of more than 90 years.

WILDEBEESTS

The life of these African antelopes (also called gnus) is one 16-year trek. Despite their large numbers, their diet is so special-ized that they are always on the move in search of their preferred grass. They also require more water than many other African beasts, but have developed a recycling process for their body water that enables them to survive dry spells.

WOLVES

In captivity, wolves can live up to 16 years. In the wild, their spans are significantly shorter. Wolves originated as a species roughly 7 million years ago and spread across North America almost immediately. How much longer they will survive in this country is in doubt, as ranchers and hunters have eliminated them from all but small and shrinking sections of the northern United States.

WOMBATS

The coarse-haired wombat resembles a small bear and can live more than 20 years in captivity.

WOODCHUCKS

Also called groundhogs, woodchucks are among the animals that have benefited from man's encroaching on nature. They prefer farms to forests and live up to 15 years if they escape the gardener's wrath.

ZEBRAS

The convicts of the equine world, zebras live about 25 to 30 years, but have reached 40 in captivity.

Birds

Birds first made their appearance on earth during the Jurassic period, 180 million years ago. The crucial innovation that gave them a competitive edge over their immediate ancestors, the flying reptiles, was their feathers. Between a bird's feathers and skin a layer of air can be trapped, effectively insulating its body from the vagaries of external temperatures. By changing the position of their feathers, birds control the exchange of heat between their bodies and the surrounding air. With this advantage they advanced up the evolutionary scale as warm-blooded animals active in all weather.

By the end of the Miocene era, there were approximately 11,000 species of birds. Since then, the number of species has declined because of changes in the earth's climate and vegetation as well as the damaging effects of man and civilization. Today there are still about 8,600 species of birds inhabiting the world. In the struggle for survival they adopt one of two strategies, depending on the general characteristics of their habitats. In unstable environments—for example, where land is still in the process of being overgrown by trees—birds attempt one large reproductive effort early in life. These opportunists are generally small birds that lay large clutches of eggs and tend to die young. Fortunately, they can adapt well to different habitats and feed on a wide variety of foods. In stable habitats in which food supply remains fairly constant live large birds that breed several times, but whose reproductive lives begin relatively late. Because these species have low reproductive rates, they have adapted for survival: Their large size protects them against predators; they care for their young for a prolonged period of time; and they are extremely efficient at finding food, although their diets may be more specialized than those of their short-lived relatives. However, while these very attributes extend their life-spans, they also tend to restrict the population size of such birds, as well as the birds' ability to adapt to rapidly changing environments. Hence they are vulnerable to extinction.

It should come as no surprise then that large birds such as the California condor, the North American whooping crane, and the New Zealand kakapo are among the most endangered bird species. Those large birds that continue to thrive, however—parrots and swans, for example—enjoy life-spans that may exceed 70 years.

Birds that have involuntarily relinquished their freedom for protective internment in cages or zoos do not necessarily enjoy longer life-spans in return. Specimens caught in the wild generally do not take kindly to capture and frequently die in transit. A notorious example is the beautiful cock of the rock; for every such bird that survives transport down from the Andes, 50 others perish.

BLUE JAYS

Relatives of crows, magpies, and ravens, blue jays live up to 14 years in the wild. A hungry blue jay poses a fatal threat to other birds: If it isn't satisfied with a meal of insects, seeds, and nuts, it eats other birds' eggs and nestlings.

BUDGERIGARS

Often called the "budgie," this small Australian bird is the best known of the true parakeets and has a life-span of 6 to 15 years. Its numbers fluctuate wildly as the food supply of seeds and foliage flourishes or declines. Remarkably fertile, a male budgie is fully mature within 60 days of fledging.

CANARIES

The *Guinness Book of World Records* reports a 31-year-old canary, but canaries over 20 are extremely rare.

CARDINALS

The record of longevity for cardinals is 30 years. Cardinal marriages also are long-lasting; the males not only captivate the females with their serenades, but are also generally monogamous and help rear their young.

CHICKADEES

Chickadees live up to 10 years in the wild.

CHICKENS

The domestic chicken has a life-span of 8 to 14 years—but only if it is not marked for the pot early in life.

COCKATOOS

A sulfur-crested cockatoo has survived 56 years in captivity. One was actually reported to have lived 120 years, although the report could not be proved.

CONDORS

Condors, the common name given to various vultures, are the largest living flying birds. Weighing up to 30 pounds, the Andean condor has lived more than 60 years in captivity, and one in the Moscow Zoo lived to be 72. A rare California condor, slightly larger than its South American cousin, has lived 37 years in captivity. Condors lay only one egg every other year and do not breed until age 5. It takes a pair of condors an estimated 21 years to be certain of raising two chicks to replace themselves; not surprisingly, the species is near extinction.

CORMORANTS

One of these sea birds, banded as a chick, lived more than 13 years.

CRANES

A white Asiatic crane has lived more than 60 years in the National Zoo in Washington, D.C. Others in the crane family have been known to live well into their forties.

CURLEWS

There are instances of banded wild curlews living for more than 30 years. Another testimony to their hardiness: Curlews can fly nonstop for more than 2,000 miles.

DOVES

Doves are actually pigeons (of which there are 289 species) and live up to 30 years in captivity. Their usefulness to man is legendary. From the time of Noah they were trained to carry messages—and to race. Symbolizing peace, gentleness, and love, they were sacred to Aphrodite. Today they still are valued as game birds and for use in scientific experimentation.

EAGLES

One of the smaller eagles, the bataleur eagle of Africa, has lived 55 years in captivity. There are examples of golden eagles living more than 40 years, crowned eagles, more than 30 years. The eagle also is an ancient and venerable symbol. It served as the emblem of one of the Egyptian Ptolemies; it later emblazoned the standards of both the Roman and the Napoleonic armies; and in 1782 it was voted the symbol of the United States.

GEESE

Geese are the longest-lived domesticated birds, surviving about 25 years. They are often force-fed to enlarge their livers, which are used to make pâté de foie gras.

GRACKLES

A cannibalistic nest robber, the grackle lives up to 15 years.

GULLS

Gulls live for more than 30 years in the wild. Several species have shown spectacular population increases in recent years, probably as a result of the vast quantities of human garbage added to their food supplies.

HAWKS

Hawks generally survive less than 15 years. Ospreys, harmless brown hawks, have been known to pass the 20-year mark. In recent decades a dramatic decline in the numbers of several species both of hawks and of falcons, their close relations, was noted in America and Europe. Derek Ratcliffe, a British ornithologist, observed that many peregrine falcon nests contained broken eggs and that parents often devoured the contents. He discovered that DDT interferes with the birds' hormonal systems controlling the calcium deposits in eggshells. As a result, the thin-shelled eggs were crushed during incubation. Both birds of prey and predatory species that live off marine life are particularly susceptible to pesticide poisoning.

HUMMINGBIRDS

These tiny creatures are among the shortest-lived birds. The longevity record for hummingbirds was set by an emerald-throated hummingbird that lived over 10 years at the Bronx Zoo.

MOCKINGBIRDS

The preeminent songster of all North American birds, the mockingbird survives a little over 10 years in the wild.

MYNAS

The best talker among the birds, possessing exceptional talents for mimicry, the myna can live up to 20 years.

OSTRICHES

The largest of living birds, the ostrich has a life-span of about 25 years, although one aged 62 was killed in South America in 1972. During its lifetime, the polygamous male is kept busy tending his flock of up to six females during the day and incubating their eggs (nearly 3 pounds apiece) at night.

OWLS

Owls generally have few natural predators, but they do run the risk of being mobbed by gangs of smaller birds. A giant European eagle owl lived to be 68 years old at the Bronx Zoo. By contrast, the laughing owl of New Zealand is now extinct; as is true of many ground-nesting birds, its decline was attributable to predation by alien mammals introduced by European settlers.

PARROTS

Parrots, of which there are 315 species, live up to 50 years, although some have made it past 70.

PELICANS

The life-span of a pelican in captivity is more than 30 years. One resident of the Rotterdam Zoo was thought to be at least 49 years old at its death.

PENGUINS

The emperor penguin, the largest of these regal birds, stands about 4 feet tall. An emperor, however, prefers not to stand: A bird most suited to marine life, its webbed feet make better rudders than ambulatory limbs. And during the 2 months it spends on land incubating its young, storing its eggs in a fold of skin between its feet, a penguin won't eat. In captivity, penguins enjoy an average life-span of 34 years.

PIGEONS

Pigeons live up to 30 years in captivity. Although we usually associate pigeons with urban environments, the now-extinct dodo bird, last found only on the island of Mauritius, was a member of the pigeon family.

SPARROWS

Despite their small size, sparrows live more than 20 years. Their longevity is no doubt increased by their ability to feed on a wide variety of foods, including seeds, insects, and all manner of human debris.

STORKS

Storks live more than 35 years in captivity, an acceptable lifespan for the bird whose appearance is considered an omen of fertility.

SWANS

Swans live as long as the bird is beautiful—up to 70 years.

SWIFTS

Swifts, the fastest birds alive, live only about 12 years.

VULTURES

These close relatives of the eagles have long wings that allow them to remain airborne effortlessly for long periods as they scour the countryside for carrion over the 35 or more years of their lives. Because there are no longer sufficient animal corpses for them to eat, several European species are in serious decline. In an effort to prevent their extinction, the Spaniards have begun leaving carrion in remote "vulture restaurants." Instead of posing a health hazard, this practice may actually be a hygienic method of disposing of animal carcasses: The digestive juices in a vulture's intestines destroy harmful bacteria that might otherwise find their way into the water supply.

WOODPECKERS

Woodpeckers live an average of 10 years. Scientists have yet to determine exactly how much wood a woodpecker pecks during its lifetime. Rather than peck more, smaller and less powerful woodpeckers restrict their efforts to softer trees in more advanced stages of decay. The population of one of the most exquisite American birds, the ivory-billed woodpecker, has declined along with the timber industry. In 1968 only six breeding birds were known to exist.

Fishes

Determining a fish's age or calculating the average life-span of a particular fish species poses a unique set of problems. Fishes are difficult to observe, since they exist in a habitat thoroughly inhospitable to man. Although ichthyologists tag fishes, it is virtually impossible to observe them closely in the wild over lengthy periods of time; yet in captivity, where problems of predation and competition do not occur, fishes may far exceed their natural life expectancies. And then there is the problem of numbers: Not only are there over 20,000 species of bony fishes alone, but life-spans vary as well among the subspecies or races that make up a species. Also, growth rates—and hence life-spans—often differ widely as a result of a population's particular habitat. Variations in the food supply, the weather, and the water's temperature and chemical composition all play a part in whether a fish will attain its genetic potential. (In colder waters, for example, a fish's metabolism slows considerably; not only may it grow and sexually mature more slowly, it is also likely to live longer than its southern relatives.) Finally, there is the tremendous range in life-expectancy figures. Some fishes are annuals—minnows and gobies rarely live past the first year; others, such as salmon, pike, and sturgeon, are credited with exceptionally long life-spans rivaling, and perhaps exceeding, man's.

Fortunately, a fish's anatomy provides the measures by which to estimate its age. The most scientific and accurate method involves an examination of the fish's scales. Scales— whose number remains constant—develop growth rings as they increase in size. During winter, or when the fish spawns, the scales' rate of growth slows and the rings occur so closely together that they form dark bands or annuli. A trained ichthyologist can determine a fish's age from the number of annuli on the scales.

A simpler way of assessing a fish's approximate age is by measuring its size. Theoretically, a fish is capable of indefinite growth, although the growth rate decreases as the fish ages. Therefore, the larger the fish in comparison with its schoolmates, the older it is. However, for the reasons mentioned above, no table can be drawn relating size and weight to exact age for each particular species. For certain types of fishes, though, size serves as a suitable yardstick for measuring age.

AMERICAN EELS

Although gluttonous feeders, American eels (only catadromous species) grow slowly, travel thousands of miles to spawn, and die after a single spawning. They routinely live for 7 to 20 years, depending on species and location, but have survived up to 60 years in captivity.

BARRED SURFPERCH

The barred surfperch is such an important catch to California anglers that the state's Department of Fish and Game conducted an intensive study of its life cycle. The statistics compiled indicated that females live up to 9 years, but no male specimens over 6 years old were found.

BLACKBANDED SUNFISH

This little sunfish has a maximum age of 4 in the wild, but can survive up to 6 years in an aquarium.

BROOK SILVERSIDES

Also called skipjacks, these small, slender fish are among the annuals; few survive a second winter.

CALIFORNIA YELLOWTAILS

Because of popular interest in the yellowtail, its life history has been studied in detail. Tagging experiments have yielded significant data on the fish's movement patterns, and analysis of scale annuli indicates that these fish live to more than 12 years of age, at which point they may be more than 45 inches long and weigh about 27 pounds.

CARP

This hardy—and smart—fish has been found in abundance in Europe since earliest times (Aristotle even wrote about it). A prolific breeder, it has become widely distributed in North American waters since its introduction. Carp are able to survive under a wide range of conditions. They tolerate low oxygen levels and extreme variations in temperature (they can withstand water temperatures to 96° F for 24 hours as well as temporary freezing). No wonder then that there are legends of carp living to 400 years of age. Their actual life-spans, however, are 20 to 25 years, although they have lived up to 47 years in captivity.

CATFISH

There are at least 15 families of catfish scattered throughout the world, many of them highly specialized and distinctive: There are walking catfish, talking catfish, blind catfish, climbing catfish, and parasitic catfish. Catfish taken in Europe have had life-spans of up to 60 years.

COD

There are about 60 species of cod (haddock and hake are included in this fish family), all marine fishes of cold-temperate or arctic waters. Although they live to a respectable 15 years of age, their infant mortality rate is astounding: Although the female cod can lay as many as 5 million eggs in a single spawning, only half a dozen or so survive.

DOLLY VARDEN

Named after a character in Charles Dickens's *Barnaby Rudge* (they both dress in pink-spotted attire), this western char is very inactive for the first 4 years of its life. It grows slowly, doesn't reach maturity until age 6, and lives more than 18 years.

DOLPHINS

Dolphins (the fish, not the porpoises) have extraordinary growth rates—5 pounds a month—but a short life-span of 3 to 4 years.

FLOUNDERS

There are more than 200 species of these flatfishes, all having in common a broad, flat body and the location of both eyes on one side of the head. The American plaice has been known to survive past age 26; the Dover sole lives to at least 15 years; the Pacific sanddab—a lefteye flatfish—lives at least to 10; but the hogchoker makes it only to 7. Pacific halibut are actually members of the righteye flounder family. Growing to 9 feet (500 pounds), they live to at least 35 years of age, during which time their migrations and peregrinations have taken them at least 2,000 miles.

GIANT SEA BASS

This huge marine bass has been known to live to 75 years of age, attaining a weight of more than 550 pounds.

GOBIES

Gobies are among the shortest-lived vertebrates. They are born, reproduce, and die within a single year.

GOLDFISH

Related to the carp, goldfish originally came from China. A life-span of 7 years is usually considered admirable, but there have been goldfish that lived to 30 years of age.

GUPPIES

Guppies have been known to live up to 6 years.

HADDOCK

Haddock are found in deeper Atlantic waters than their cod relatives, but they share the same approximate life-span of 14 years.

HALIBUT

These members of the righteye flounder family are among the largest marine bony fishes; the Atlantic halibut, for example, can grow to a weight of 700 pounds and a length of over 9 feet. Such gargantuan size indicates a long life-span; in fact, halibut often survive past 35 years. One, caught in 1957, was estimated to be over 60; it was a female and was both fertile and still growing.

HERRINGS

Until recent times herrings were the most abundant type of seafood in the world. The anadromous American shad live for 8 or 9 years. Shad that live south of Cape Hatteras die shortly after spawning; after a long migration and a tremendous loss of body weight (they won't eat in fresh water), they simply no longer have the fat reserves to sustain a return to the ocean. Their northern cousins, however, are able to make it back to sea. Pacific herrings usually live to 8 years of age, but occasionally they survive to 20. The Pacific sardine's life expectancy averages 10 years, but it can live to at least 13. Needless to say, most sardines—the general name for young herrings—find their way into tin cans long before they reach senescence.

INCONNUS

Despite the name (meaning "unknown"), the inconnu is a highly prized predatory game fish, often reaching 55 pounds. Females mature late (at 7 to 12 years of age). While spawning occurs only every 3 or 4 years, the fish is blessed with a long life-span that allows enough time for it to reproduce sufficiently: 21-year-olds are not uncommon.

MINNOWS

Minnow is derived from the Latin word for "small," and generally this is true of the approximately 200 species in North America alone (the notable exceptions to the rule are the carp and squawfish, which can grow to 80 pounds). Most varieties of minnow are short-lived; the dusky shiner, for example, rarely lives past 3.

MUSKELLUNGE

This prize freshwater game fish is one of the largest—and fastest growing—members of the pike family. Although its usual life expectancy is only 3 to 6 years, there are several records of 18- and 19-year-old fish. The oldest recorded muskellunge was estimated to be 30 years old and weighed 70 pounds.

OYSTER TOADFISH

The life expectancy for a male oyster toadfish is 12 years; the female, however, averages a mere 7 years.

PADDLEFISH

A still existent member of one of the older groups of fossil fish, the paddlefish not only can look back on a long history but also can look forward to a long life. Specimens up to 30 years old have been taken in Lake of the Ozarks, Missouri.

PERCH

The perch family includes darters, saugers, and walleyes as well as a wide variety of perch. The yellow perch is the most widely distributed and has a life-span of up to 11 years in northern waters, 5 or 6 years in southern rivers. But the odds that this slow swimmer and ideal forage fish will survive its first year are 1 in 5,000. The white perch has a somewhat longer life-span: 12-year-olds are common and at least one 17-year-old has been recorded. (White perch are actually members of the bass family.) Ocean perch, despite the strong likelihood that they will end up as frozen fish sticks, fare much better. Slow-growing and maturing at age 10, they generally survive to age 20, and some specimens have reached 27 years of age.

PIKE

One of the fastest-growing fish, these ravenous freshwater counterparts of sharks have been called "mere machines for the assimilation of other organisms." Few species of fish have inspired as many fables and superstitions as the pike. The famous Mannheim Pike was supposed to have been 267 years old. Actually the average life-span is close to 10 years for the chain pickerel as well as northern pike (although individuals live to 20 or more).

QUILLBACKS

Although a quillback's life-span is estimated at 8 years, the species is ideal forage for other predatory fish, and the mortality rate is high. Few quillbacks live out their natural life-spans.

SALMON

Salmon actually belong to the trout family, along with white fish, grayling, and char. The Atlantic salmon has a life-span of more than 10 years, spawning at about 3 years of age. The sockeye salmon—one of the most commercially valuable— lives to age 8. The Pacific salmon's life expectancy falls between 2 and 8 years, depending on when it spawns, since it expires soon afterward.

SHARKS

There are over 250 species of sharks, ranging from 2 to 60 feet in size. A very old group of fish, they lack true bone cells and have skeletons made up entirely of cartilage. Shark life-spans are difficult to determine because few of the larger sharks do well in captivity (and who wants to get close enough to tag one?). However, a nurse shark about 14 feet long lived for 25 years in the Shedd Aquarium in Philadelphia.

STRIPED BASS

The age of this anadromous fish can be determined by its size: A 2-year-old is 12 inches long and weighs 0.75 pound; a 10-year-old, 38 inches, 30 pounds; a 14-year-old, 41 inches, 40 pounds; and an 18-year-old, 50 inches, 50 pounds. At least one 23-year-old striped bass has been recorded.

STURGEONS

Sturgeons are primitive fish that were widely distributed early in geological time throughout the globe. They are slow-growing, maturing in 12 to 22 years, depending on species, but living 75 years or more. The record for sturgeon longevity belongs to a 13-foot-long beluga sturgeon that weighed about 1 ton and was thought to be 82 years old. A lake sturgeon caught in 1953, however, was believed by some experts to be around 150 years old.

TROUT

The longevity of these salmonids varies according to species. Cutthroat trout—despite their name—do not compete well with other fish and live only 6 to 9 years. The life expectancy for rainbow trout falls between 7 and 11 years, depending on race and locality. Brook trout, the finest trout at table, do best in cold waters: In some Canadian waters they live up to 10 years; elsewhere a 6-year-old is considered senescent. One brook trout in captivity lived up to 18 years.

TUNA

Aristotle wrote that the tuna grows to a weight of 1,200 pounds by age 2 and then dies, an inaccurate estimate to be sure. But this member of the mackerel family does grow rapidly, and bluefin tuna have been known to exceed 1,500 pounds. Fourteen years is considered a long life-span for a tuna.

WALLEYES

The largest member of the perch family, the walleye has a life-span of 6 to 7 years in southern states, but its life expectancy doubles in northern waters. A relative of the walleye, the sauger, inhabits only the largest rivers and bodies of water. A Canadian or northern sauger lives for 10 to 14 years; its confederate cousin survives only 5 to 7 years.

WHITE BASS

The average life-span for the white bass is 3 to 4 years in warm waters, 4 to 5 years up north; maximum life-spans in warm and cool waters are 6 to 7 and 7 to 9 years, respectively.

WHITEFISH

Extremely slow-growing, whitefish live up to 16 years, attaining lengths of 25 inches.

Amphibians and Reptiles

In his study of longevity in 1798, W. C. Hufeland noted: "Amphibia, those cold and doubtful beings, can prolong their existence to an extraordinary length." Indeed they can, for these relatively primitive creatures often reach 40 years of age, regardless of their size and period of growth. Furthermore, they are among the oldest species in the animal kingdom. The first successful land vertebrates were ancient amphibians—the labyrinthodonts—that closely resembled their ancestral lobe-finned fishes, but had evolved limbs strong enough to support their body weight on land. Most of these primitive creatures disappeared by the first part of the Mesozoic era, 225 million years ago. Those that survived were the precursors of the modern amphibians—the frogs and salamanders—or had evolved one step further along the evolutionary path to become the earliest reptiles—the colylosaurs or stem reptiles. Appearing about 250 million years ago, these were the first vertebrates capable of existing entirely on land (amphibians must return to water to reproduce). In turn, they evolved into a great variety

of descendants—turtles, lizards, snakes, alligators, and croco-
diles, as well as the flying and giant "ruling" reptiles that disappeared in a flash extinction roughly 70 million years ago.
The most famous of the giant reptiles were, of course, the di-
nosaurs (meaning "terrible reptiles"). Despite the disappearance of numerous species, more than 2,000 species of
amphibians and 6,000 species of reptiles survive today. One,
New Zealand's tuatara—the only remaining rhynchocepha-
lian—similar to an alligator, has existed as a species for 200
million years, living proof of the longevity of the amphibians
as a class of animals.

ALLIGATORS

There are two species of alligators, the large American type
and the small Chinese one. One Chinese alligator lived over 52
years in captivity. Now protected by law, the American alligator should have an easier time of it; the Chinese variety, however, is nearly extinct.

CROCODILES

Crocs, one of the oldest extant species of reptiles, live long
lives. One made it through 56 years of captivity.

FROGS

The longest-lived of the toads and frogs is the common toad,
which has lived as long as 36 years in captivity. The common
bullfrog and the tree frog live up to 16 years. In South America, the local version of the bullfrog survives 12 years, the
arboreal frog, up to 6 years.

GECKOS

These nocturnal lizards—the smallest reptiles—live up to 7
years.

GILA MONSTERS

Gila monsters survive up to 20 years in captivity.

IGUANAS

These are the tiny reptiles sold in pet shops. They can live up to 25 years, by which time they will be about 6 feet in length. The land iguana, found only on the Galápagos islands, has a life-span of about 15 years.

LIZARDS

Lizards are the most abundant of living reptiles. There are some 3,000 species, ranging in length from about 2 inches to over 12 feet. The anole lizard, the species most widely kept in home terrariums, has a life-span of about 4 years. The longest-lived lizard on record belonged to the species known as the limbless lizards, or slow worms; it survived 54 years in captivity.

SALAMANDERS

Like lizards, salamanders have a wide range of life-spans; several species survive up to 25 years. The record, however, belongs to the Japanese giant salamander, one of which lived at the Amsterdam Zoological Gardens for 52 years and was thought to be about 55 years old at the time of its death in 1881. It grew to be 5 feet long.

SNAKES

The common garter snake has a life-span of 6 to 10 years. Larger snakes—such as the anaconda, cobra, and leopard snake—can live up to 30 years. A boa constrictor living in the Philadelphia Zoo is already over 41 years old.

TORTOISES

Turtles and tortoises belong to the same order (Testudinata). The difference is that tortoises are primarily terrestrial creatures, while turtles are primarily aquatic. After the sea anemone and the sponge, the black Seychelles tortoise holds the record for the longest life-span of all animals: 170+ years. Several other tortoises live over 100 years, among them the giant tortoise of the Galápagos, the European pond tortoise, and the Moorish tortoise. Not all centenarian tortoises are large; the Moorish tortoise, for example, is only about 10 inches across.

TURTLES

Only about 6 inches across, the Carolina box turtle is the longest-lived of the turtles. One of these little turtles was marked, set free, and subsequently retrieved—129 years later. Snapping turtles have been kept up to 59 years in captivity; the terrapin, or North American musk turtle, lives more than 50 years; the mud turtle, up to 25 years; and the North American spotted turtle, up to 42 years.

Invertebrates

Invertebrates are extremely inconsistent when it comes to life-spans. They range from among the shortest-lived (single-celled animals that live only hours) to among the longest-lived (the sponges and sea anemones, which survive so long that they seem virtually immortal).

Studies of various invertebrates have found what seems to be an inverse relationship between the degree of cell replacement and the susceptibility to senile change; that is, the more often an invertebrate changes its cells, the less susceptible it is to aging.

ANTS

Most ant colonies contain three castes—queen, males, and workers (wingless, sterile females). Queens can live up to 15 years; workers, up to 7 years. The males die soon after mating, however, and their life-spans may be as short as a few days.

BEES

A queen bee may live up to 6 years; a worker bee, about 6 months; and drones (males), about 8 weeks.

BEETLES

Beetles live from 1 to 3 years.

CLAMS

The longest-lived invertebrate, the European freshwater clam, is thought to live up to 116 years. Clams found in U.S. waters may live up to 60 years, and the giant clam (whose shell alone sometimes weighs over 500 pounds), about 30 years.

COCKROACHES

There are more than 3,500 living cockroach species, of which the American and the Oriental are the most common. The large black cockroach often seen in basements and around apartment buildings is the Oriental cockroach, with a life-span of about 40 days. The American cockroach, a red or dark brown insect that can be as long as 2 inches, has a life-span of over 200 days. While the life-span of an individual cockroach is not all that spectacular (a year at most), as a group, roaches cover an impressive stretch of history. Fossil cockroaches, almost identical to the ones that exist today, date back 250 million years.

CRABS

There are thousands of species of crabs. They range in size from those that are as small as a fingernail to others that are 6 or 7 feet long, stand 3 feet high, and weigh up to 40 pounds. Life-spans of some of the better known crabs are: hermit crabs (which get their name from occupying shells abandoned by other crustaceans), 11 to 12 years; blue crabs, 2 to 3 years; Dungeness crabs (a West Coast variety), 8 years; Alaskan king crabs (which grow up to 24 pounds), over 15 years.

CRICKETS

The common field cricket has a life-span that runs from 9 to 14 weeks—if it is lucky. Crickets make their distinctive sounds by rubbing their wings against a hard ridge on either side of their bodies. Usually it is the right wing (not only are left-wing crickets in the minority, but they are not as loud as right-wing crickets). In the scheme of cosmic events, the death of one cricket is usually not given much attention, but in the more decadent days of China it was. There, cricket fights were high-stake duels to the death. The insects were placed facing each other in a small pit, and the referee, using an ivory wand with a whisker attached to its end, would irritate the crickets' antennae. The animals would then rush at each other and try to tear each other limb from limb. Whichever survived, won. Star crickets could sell for as much as $100, and there is mention of one cricket who made as much as $90,000 for its owner. When this champion died, it was given the honorable title of Victorious Cricket and was buried in a tiny silver coffin.

EARTHWORMS

These invertebrates, vital to agriculture, live from 5 to 10 years.

FIREFLIES

It is difficult to pinpoint precise life-spans for fireflies, since about 1,300 different species are distributed around the world. Normally, adult fireflies of most of these species live only a few weeks. But they spend between 2 and 3 years growing to adulthood.

FLIES

The common housefly, *Musca domestica*, is only one of 85,000 fly species. The average life-span for both sexes runs between 19 and 30 days, male life-spans being generally in the lower range. There is record of an extremely sheltered fly that lived to be 70 days old. In her lifetime the average female lays two to seven batches of 100 to 150 eggs, most of which do not survive to adulthood. This is just as well, since according to one Chinese study a housefly raised in a *clean* neighborhood carries on its hairy body close to 2 million bacteria, while a fly from a dirty part of town may have as many as 3.5 million bacteria on its back. By one entomologist's computations, if all the offspring of a normally fertile fly couple born between April and August survived and then died all at once, the earth would suddenly be covered with 191,010,000,000,000,000,000 fly corpses peacefully resting three stories deep.

GRASSHOPPERS

There are almost 9,000 different species of grasshoppers. The familiar green insect we know by that name lives only about 5 months, whether in the wild or in captivity. The female katydid fares somewhat better. From egg to adulthood takes her about 2 years. But from then until death, her life-span is brief. She lives only a few weeks—long enough to fertilize and lay her eggs—and then dies.

LOBSTERS

Lobsters are the longest-lived crustaceans. A 35-pound lobster is estimated to be about 50 years old, and the largest lobster on record, caught off Long Island and weighing 44.5 pounds, was probably even older. A 1.5-pound lobster, the popular eating size, is about 8 years old.

LOCUSTS

Locusts actually belong to the grasshopper family. The desert locust, which migrates in vast herds across Africa and Asia and is thought to be the locust of Biblical fame, has a life-span of about 75 days.

MAYFLIES

Although they have a reputation for impermanence, mayflies may live as long as 1 to 3 years in the larval stage before they blossom to full adulthood as flies. Adulthood is when they begin to fade fast. With only vestigial mouths and stomachs, adult mayflies have no time for eating—they have only 7 or 8 hours left in their lives to mate, lay their eggs, and then die.

MOSQUITOES

Life is hard for the mosquito. Every 24 hours 10 to 20 percent of all mosquitoes hatched are instantly killed off. Those that make it past the first day live no longer than 2 months in the wild, although mosquitoes have been coddled and nurtured in the laboratory for as long as 5 months.

OCTOPUSES

Octopuses are mollusks related to snails, clams, and oysters. The Mediterranean octopus has a maximum life-span of about 10 years. If necessary, an octopus will feed upon other octopuses, and occasionally will even eat its own arms just prior to death.

OYSTERS

The maximum recorded life-span for a freshwater oyster is 80 years. Saltwater oysters live only up to 10 years.

SEA ANEMONES

Thought to be able to live on indefinitely, the sea anemone is considered one of the immortals. Colonies have lived nearly 100 years in captivity without signs of deterioration.

SILKWORMS

Although most celibate invertebrates live longer than those that have mated, silkworms are the exception to the rule. An unmated silkworm will live 12 days, compared with 14 days for its sexually active sister. A male that has mated lives for 15 days, extending its life-span by 1 day.

SNAILS

Land snails live up to 5 years.

SPIDERS

The maximum life-span for most spiders is 4 to 7 years. Tarantulas, however, have been known to live as long as 20 years.

SPONGES

Thought to be theoretically immortal, sponges have no known life-spans.

TARDIGRADES

Death, or apparent death, is a sometime thing for these small, beetlelike creatures that measure about .04 inch long and are believed to be arachnids, part of the same family as spiders and scorpions. What makes them items of curiosity for life-span study is that tardigrades are anabiotic; that is, they have the ability to appear to die and then spring back to life a number of times.

Deprived of all food and water, a tardigrade will shrivel into what looks like a hard grain of sand and be, to all appearances, dead. Given a couple drops of water, the tiny creature will swell back up to its natural proportions and resume its normal life. Biologists figure that the tardigrade can, in its lifetime, successfully put its body through 14 of these life-"death" anabiotic cycles, some of which can drop a tardigrade into its special state of suspended animation for as long as 6.5 years. The maximum awake-and-moving life-span for a tardigrade is 30 months, but, using its anabiotic cycles, a particularly tenacious one could extend its life-span to as long as 67 years.

TERMITES

An individual termite worker can live about 20 years; queens, the longest-lived members of the species, have been known to survive for 50 years—still laying eggs. Not only do termites have extraordinarily long life-spans, these insects also leave spectacular architectural monuments behind them when they eventually expire. The termite mounds that dot the savannas of East Africa, for example, are insect-scale versions of the twin towers of the World Trade Center. Every rainy season millions of termites leave an established colony and strike out on their own. The lucky ones that aren't eaten by waiting birds or aardvarks settle down a few hundred yards away from the mother colony and build a small nest chamber in the ground. Over the years the offspring multiply and their home grows right along with them. Every mound is an intricate maze of tunnels with cunningly made vertical air shafts, often reaching 20 to 30 feet in width and 6 to 10 feet in height. Each is made up of loose dirt cemented by termite saliva, a building material so hard that scientists have broken steel picks trying to dig into a mound. In about 30 years time the termite colony that built the towering structure dies off, leaving behind a magnificent termite ghost town.

Microscopic Life

DNA

Although it is not in itself an organism, DNA is the fundamental building block upon which all life is based. When the first form of life appeared in the earth's primordial slime about 3 billion years ago (formed out of elements in shallow ocean water exposed to solar radiation), it probably carried a blueprint for creating others exactly like itself—in the guise of the molecule DNA (deoxyribonucleic acid), the substance genes are made of.

These first forms of life were tiny, primitive single cells, far different from the more complex single-celled animals we know today. In time, they became more complex, began to differentiate, and created the living organisms in our world. Nevertheless, the process was a slow, logical one, in which DNA was constantly passed from one generation to the next, from one species to the next, from one family to the next. As a result, it can be claimed that every living thing contains some elements of that original DNA and that DNA has thrived and survived for all these 3 billion years.

VIRUSES

Viruses consist of nothing more than squiggles of DNA—the hereditary material—wrapped in coats of protein. Their sole function in life is to reproduce, which they do by inserting themselves into larger organisms, such as bacteria or body cells, and borrowing the reproductive mechanisms that are already available there. They are, in short, parasites.

The life-span of a virus—and, for that matter, of any single-celled animal (ameba, paramecium) that reproduces asexually—is a bit elusive. Theoretically, a virus can live forever, given the right conditions. It simply divides, generates whatever it lost during the division, grows, and divides again, with each division creating two viruses identical to the first. It is, therefore, more meaningful to talk of a virus's *life cycle*—the time it lives between divisions. For most viruses, that cycle lasts only a few hours.

The problem with viruses is that they do not always act like living things. One University of California virologist managed to isolate the tobacco mosaic virus in a crystalline—and apparently dead—form and was able to bring it back to life by dissolving it in a nutrient solution. In a sense then, viruses are our microscopic ghouls.

BACTERIA

Bacteria are among the simplest, hardiest, and, some biologists believe, earliest forms of life. They are responsible for everything from fermenting wine to helping us digest our food. (The General Electric Company, through sophisticated genetic engineering, designed its own unique, patented bacterium that eats oil—breaking it down into water and carbon dioxide.) Bacteria thrive on inorganic material, and many can do quite well without any oxygen. In fact, to some strains, oxygen is a poisonous gas.

The life-spans of bacteria constitute an open-ended scientific question. Bacteria have been found still alive in 10- to 20-million-year-old rocks dredged up from the floor of the Atlantic Ocean. They were discovered by marine biologists of the U.S. Geological Survey, who theorized that the bacteria live on organic matter and oxygen trapped in the rocks.

What may be the oldest revivable bacteria were discovered by a West German scientist, Dr. Heinz Dombrowski, in prehistoric salt deposits estimated to be about 650 million years old. He extracted the bacteria from the deposits and placed them in special nutrient solutions, where the ancient bacteria perked right up and came back to life. He identified them as *Bacillus circulans*, describing them as bacteria that thrive in old milk cartons and dirty milk bottles.

The life *cycles* of bacteria are a bit easier to measure. They range from several hours for some species to 20 minutes for *Escherichia coli*, the tiny symbiont that colonizes our own intestines.

Preserved Animals

To find animal specimens that were as deliberately and as carefully preserved as those of today's taxidermists, you have to go back to the ancient Egyptians. Archeologists have found the mummified remains of everything from cats to baboons secreted in the tombs of the royalty of ancient Egypt. One of the oldest and oddest finds was the body of a young female hamadryad baboon embalmed and placed inside a royal coffin next to that occupied by Queen Makeri. The well-preserved corpse of the animal, over 3,000 years old, was identified on the outside of the coffin as Princess Moutemhit, the queen's daughter.

Modern taxidermy began in 1920, when a taxidermist decided to skin an animal and cast a body mannequin in clay, before tanning the skin and replacing it over the mannequin. But taxidermy itself is an ancient art: The Romans stuffed gorillas; various religious cults stuffed bear and bird heads for ornamentation; a whole rhinoceros that was stuffed during the 1600s survives today in Europe; and a lion mounted 125 years ago still resides in the American Museum of Natural History in New York City.

Unfortunately, while the products of the older methods of taxidermy were durable, they had a serious drawback: over long periods of time they began to stink. The new method, which now uses anything from clay and plaster to fiberglass to form the body, is both odor-free and longer lasting. In fact, if it is cared for correctly, today's stuffed animal can "live" indefinitely.

Imaginary Beings

The imagination of man has spawned a zoo full of bizarre and wondrous creatures. Most came equipped with life-spans longer than that of their creator.

CENTAURS

The centaur is an immortal—half man, half horse. Lucretius (the Roman) maintained that centaurs were an impossibility, reasoning thus: Since horses and humans develop and age at different rates, after 3 years the rear half would be full grown while the front half would still be a babbling baby. Centaurs are uncouth and savage; they feast on raw flesh, drink incredible amounts of booze, and chase women. But not all of them are destructive: one, Chiron, was noble. As the wisest of the centaurs, he educated Hercules and Achilles. But Hercules wounded Chiron by accident, and the pain was so great that Chiron surrendered his immortality to Prometheus and died. Zeus then set him among the stars, as the constellation Sagittarius.

DRAGONS

The life-spans of dragons have never quite been determined, since dragons seldom die of old age. Some appear to be immortal. The giant red dragon of the Book of Apocalypse in the Bible symbolized ever-present Satan for the Christians. In China, the dragon has been a national deity, used more as a model for the ideal man than as a force that must be overcome; there, every 12 years, the dragon springs to life again on the New Year as one of the symbols of the Chinese calendar.

DRYADS

More properly called hamadryads, these spirits live in trees and their life-spans coincide with those of their dwellings. Whenever a tree is felled, the dryad's death wail—too high for the human ear to hear—causes all the trees in the forest to shiver and shake.

DWARFS

Dwarfs—folklore's tiny humanoids—are gnarled little men who live in caves and in subterranean places. They are selfish and greedy hoarders of treasure and live about 200 years.

FAIRIES

Fairies are not necessarily tiny; some are as large as we are, if not larger. The Western fairy lives 9 ages and 9 ages more; the Chinese fox, a form of fairy, has the capacity to take on any shape it desires during its 1,000-year life-span. Able to cross the barriers between space and time, the race of fairies will live until Doomsday, although some will be killed and others will vanish from the earth.

Elves, who are much like fairies, are immortal, unless they choose to live with humankind. They then have life-spans about three times as long as ours, although they have no souls and perish utterly when they die. Yet " 'tis one of their tenets that nothing perisheth," according to Robert Kirk, who was carried off by fairies after he wrote a book about them.

GNOMES

Gnomes, slightly smaller than dwarfs, are tiny subterranean creatures made of earth, who are often represented as hunchbacked, red-eyed, and runty-legged. As earth spirits, they know where precious ore and diamonds are buried. Gnomes sometimes guard hidden treasure, although, more often, they just hide. They live 400 years.

HOBBITS

Hobbits, who grow to a size of about 3 feet, are smaller than dwarfs but larger than gnomes. They often live 100 years and celebrate their 111th birthdays as special occasions.

NYMPHS

These nature spirits dwell in a variety of places: The nereids live in the sea, while the naiads prefer fresh water. Dryads (see above) inhabit the trees, but the oreads sequester themselves in mountain caves. Naturally gentle, they nevertheless will attack if they feel threatened, showering magic spells and curses upon their victims. These soulless creatures do not age, but after 10,000 years they die and disappear completely.

PHOENIX

The phoenix is a large, sweet-voiced bird of gorgeous red and gold plumage. Only one lives at a time in the Arabian desert, for perhaps 500 years, or maybe 1,461 years. Then again, one example was supposed to have lived 12,994 common (Platonic) years—the time required by the sun, moon, and five planets to return to their original configuration. At the end of its life, the phoenix makes a woody nest, sits on it, and is consumed by flames. Out of the ashes crawls a white grub, which soon arises as the new phoenix.

SIRENS

These bird-women—some with mermaid tails—inhabit high cliffs near the sea. They lure unsuspecting sailors to their sides, lull them to sleep with their mysterious songs, then tear them limb from limb. But if their songs go unheeded, the sirens quickly die.

TROLLS

Trolls inhabit Scandinavian forests and caves. They come in all manner of shapes and sizes, but all will shatter if exposed to sunlight.

UNICORNS

Once found in India, unicorns were "white wild asses" with a centered horn 1.5 cubits long and colored red, white, and black. A drinking vessel made from the horn protected the drinker from poisons. (This belief was later disproved by the British Royal Society, but the cup tested was actually made from the horn of a rhinoceros, so who knows?) Later, the unicorn spread throughout the world.

Chi Lyn, the Chinese species of unicorn, eat only fallen leaves and plants already dead. Since Chi Lyn appear only when a righteous leader is born, the last one seen was in the time of Confucius. So although they live a thousand years, they are very scarce. A conflicting theory suggests that unicorns actually became extinct much earlier, when they were denied a berth in Noah's Ark by the other passengers.

3 / BEYOND THE BEASTS

FROM insects that live only a few hours to tortoises that can survive centuries—the life-spans of species in the animal kingdom seem, at first, to be hugely and richly varied. But if we look beyond the beasts, to life-spans in the plant kingdom and to longevities of mountains, continents, moon, and stars, it becomes clear that we and the rest of the animals live only a whisper in time.

Plant life alone extends the upper scale of life-spans dramatically. Scores of trees average more than a century. The bristlecone pine can survive over 3,000 years; the giant bald cypress of Mexico spans 4,000 years; and the *Macrozamia* tree, which grows in the Tambourine Mountains of Queensland, Australia, can live anywhere from 12,000 to 15,000 years. It is considered by many to be the oldest living thing on earth.

We can understand most life-spans; we can grasp, for example, the 100 years an old man might live or even the 2,000-year longevity of a redwood. But when it comes to trying to comprehend the existence of a tree that sprouted at a time when prehistoric man was still entranced by fire, we are in trouble. All of a sudden, years are barriers to understanding; the numbers we have used to try to explain the flow of time begin to lose their significance.

So what happens when we consider the astounding longevities of geologic and celestial things? When we expand the concept of life-span—that is, the length of time something exists— to include numbers in the millions and billions, as if they meant no more than the longevity of a cockroach, an elephant, or a woman in Guinea? What happens, it seems, is that we lose perspective: We may understand the numbers themselves, but we lose all sense of time. And time, after all, is what life-spans are all about.

The secret, according to those who know, is in rhythms. Our measurements of time—our seconds, hours, days, years—help us keep track. But those measurements, while they work for us, are completely artificial. They match our rhythms; they break our lives into comprehensible parts. But they are unworkable for, perhaps, the *Macrozamia* tree, to which one of our days might feel like a second, or for the sun, to which even centuries may be completely insignificant.

The numbers we are presenting are therefore only tools, guideposts to help in our understanding of these almost incredible longevities. But real understanding comes only when we try to see time as a tree or a lake or a comet would, not as *we* normally do. Let's try to do what Colin Fletcher did on his 2-month walk through the Grand Canyon, a walk he made, ultimately, to understand the delicacy of time:

I sat and rested on the rock platform, looking over and beyond the river at the strata on strata that mounted one on the other to the North Rim [of the canyon]. I could see them all, every layer. They were replicas of those I had just moved through. And after I had sat and looked at them for a while I saw that now, from a distance, I could see with my eye and intellect what I had all day been understanding through instinct. Now, as my eye traveled downward from the Rim, it watched the rocks grow older. . . .

I saw, when I looked up at the Rim, that the uppermost layers of rock were bright and bold and youthful. Their unseamed faces shown pink or white or suntan-brown, untouched by the upheavals that time brings to all of us. But below the Redwall they began to show their age. There, in staid maturity, they wore dark greens and subdued browns. And their faces had begun to wrinkle. Then, as my eye reached the lip of the Inner Gorge, the rocks plunged into old age. Now they wore gray and sober black. The wrinkles had deepened. And their features had twisted beneath the terrible weight of the years. Old age had come to them, just as it comes in the end to all of us who live long enough.

I rested on the rock platform for an hour. Then I clambered down to the river through the darkest and most twisted rock of all. Once more ... every boulder and hanging fragment of rock around me looked ready to come crashing down at any minute. But now I needed no tight and determined thinking to ward off fear. During my three weeks among crumbling rockfaces and loose talus, all apparently waiting to crash headlong at any minute, I had heard just once—a long way off—the sound of a small stone falling a very short distance. And now I understood why.

The poised boulders and fragments were indeed waiting to crash down at any minute. But there was not really too much danger that one would hit me during that particular hiccup of time we humans called May 1963. For our human clocks and the geologic clock kept different times. "Any minute now," geologic time, meant only that several fragments of rock might fall before May 2063, and that quite an appreciable number would probably do so by May 11963. I knew this now, through and through. I might not yet understand the explicit, absolute meaning of two hundred million years. But I had come to grips with the kind of geology I had hoped to find. I had begun at last to hear the rhythm of the rock.*

Celestial Bodies and Phenomena

THE UNIVERSE

According to the latest estimates, based on studies of the rate of decay of the element rhenium 187, the universe is now about 20 billion years old. Most astronomers believe the universe began in a tremendous "big bang" and has been steadily expanding ever since from the center of that explosion. Some scientists feel that the expansion will eventually stop, and then

*—Colin Fletcher, *The Man Who Walked Through Time.* New York: Random House, 1968, pp. 107–108.

a long period of contraction will begin. By this theory the universe has a finite life-span, upwards of 40 billion years. The most recent theories, however, argue that the expansion will continue forever, giving the universe an infinite life-span.

GALAXIES

The galaxies, our own Milky Way included, were formed 10 million to 100 million years after the explosion that marked the beginning of the universe. Barring unforeseen events, the galaxies should last as long as the universe itself.

STARS

Every star's life history begins with its condensation out of interstellar gas—mostly hydrogen—and ends, sometimes catastrophically, when it has exhausted its nuclear fuel or can no longer maintain its stable configuration. Gestation—the period of condensation and rising temperature—itself lasts millions of years. A star is born when the protostar's temperature becomes sufficiently high for thermonuclear reaction to begin. The newborn star's life cycle then immediately jumps into a long middle age, during which time the star shines steadily as it converts its hydrogen supply into helium. As the helium content builds up, the core contracts, and the rate of nuclear reactions speeds up until the star consists of a helium-rich core surrounded by a growing envelope of cooler gas; the star has become a red giant. When the temperature exceeds 100 million degrees, the helium begins to burn and the star shrinks in size and luminosity, and becomes unstable. It may then become an exploding nova or supernova or a pulsating variable star. A senile star is called a white dwarf; as a white dwarf it glows feebly for a few billion years more before becoming a black dwarf—a totally dead star. But if the star is too massive to stabilize into white dwarfdom, it either changes into a neutron star, known as a pulsar, or totally collapses.

The actual time it takes for an individual star to complete its life cycle depends mainly on its size and luminosity. As a rule, the largest and brightest stars have the shortest life-spans because they deplete their hydrogen most rapidly. Large bright stars live on the order of 10^6 years; thus their life-spans measure in millions of years. Medium-sized stars like our own sun last on the order of 10^{10} (tens of billions) years. Small, dim stars have life-spans on the order of 10^{13} (tens of trillions) years.

BLACK HOLES

A black hole is the burnt-out hulk of a dead star that has collapsed under its own gravitational field. The result is a space object so dense and so small that it has a gravitational field intense enough to prevent both matter and radiation from escaping. Such powerful gravitational fields have been blamed for causing wobbles in the movements of stars, and some scientists believe a black hole created the lake-sized crater in Siberia when it smashed into the earth around the turn of the century.

According to Cambridge University astrophysicist Stephen Hawking, a black hole that was formed from a star originally measuring about 1 kilometer across when the universe began about 20 billion years ago would now measure a minuscule 10^{-10} centimeter across and would be in its death throes now. As celestial objects go, Hawking says, black holes are very tenacious objects. One with the mass of our sun would be around for $10^{54} \times 20$ billion years, the present age of the universe.

NOVAS

Although the word means "new star," a nova is actually an aging star that has become unstable. Because of the resulting explosion near its surface, it increases its brightness by thousands of times. There are three types of novas—slow, fast, and recurrent. Fast novas reach maximum brightness in less than a day and then fade to their original brightness in a few months to a year. Slow novas take a month or more to reach maximum brilliancy, may remain at maximum for 10 years or more, and then return to their original brightness after decades. Recurrent novas explode in cycles of 20 to 80 years and can behave like either fast or slow novas.

SUPERNOVAS

Whereas a nova is an explosion *on* a star, a supernova is the explosion *of* a star. This fate is likely to befall all but the least massive stars as they approach their catastrophic final reckoning. At maximum, a supernova may be tens of millions of times brighter than the original star and may outshine the entire galaxy in which it explodes. Supernovas reach maximum brightness in a matter of days, then decline rapidly for about a month. The decline then slows, and the conflagration dies out over a period of years. The famous supernova of 1054, observed by Chinese and Japanese astronomers, was visible to the naked eye for 2 years. Its remnants can still be seen today and are known as the Crab nebula.

QUASARS

Quasars (an acronym derived from "quasistellar") are brilliant star-sized celestial objects that give off gigantic light or radio emissions (although their visible light is only faintly discernible through a telescope). A current theory has these space oddities as the last flashes of exploding stars out on the edge of the expanding universe—perhaps 8 billion light years away. Theoreticians now hold that the average life-span of a quasar is about 1 million years—a very long flash.

PULSARS

Among the more intriguing things to be found in space are pulsars—small, dense neutron stars that rotate in space and emit what seem like short pulses or radiowave signals. This blinking effect is similar to one you'd get by viewing a lighthouse from a distance. Pulsar waves appear to our radio telescopes as steady, predictable blinks and are so regular you could set a quartz watch by them. Depending on the pulsar, they come .033 second to 3 seconds apart.

When British astronomer Anthony Hewish first discovered pulsars in 1967, the absolute regularity of their radio pulses made him wonder whether they might in fact be artificial beacons sent out in space by some advanced civilization. In honor of that suspicion, he called the first pulsar LGM-1, for Little Green Men. Theorists have since decided that pulsars are the result of supernova explosions. Most are located from 300 to 30,000 light years from our sun. They range in age from a young 1,000 years to millions of years. As a pulsar grows senescent, its blinking becomes less and less frequent until, after about 10 million years, the blinking stops. The pulsar is finally dead.

THE SUN

The sun is a medium-sized star whose estimated life-span is 10 billion years. It is presently thought to be 4 billion to 5 billion years old.

SUNSPOTS

No one will ever get close enough to know for sure, but astronomers believe sunspots are centers of turbulence—fiery storms that move across the face of the sun. In photographs they show up as dark blotches on the face of the sun—indicating, astronomers believe, that they are about 2,000° F cooler than the sun's surface, which is estimated to be a torrid 10,000° F. The diameter of a sunspot measures from a few miles to as much as 100,000 miles. The average life-span of a sunspot is usually 1 to 4 days, but observers in 1840 recorded one that lasted 18 months.

SUNBEAMS

A sunbeam has a life-span of 8.3 minutes—the time it takes a beam of light to travel from the sun to the earth.

THE MOON

The origin of the moon is still a subject of thorny scientific debate. One theory claims that the moon was born at the same time as the earth and sun (4.5 billion years ago) from the same materials that formed the rest of the solar system. Another theory argues that it was formed in another part of the solar system and, at some point, was captured as it passed near the earth.

MOONBEAMS

Moonlight—light from the sun reflected off the moon—takes 1.3 seconds to travel from the moon to the earth.

THE EARTH AND PLANETS

Current theories place the birth of the earth and planets concurrent with, or at most a few million years after, the birth of the sun. Thus the earth and planets are 4 billion to 5 billion years old. Some scientists suggest the earth will be destroyed in the death of the sun. Others say it will outlive the sun and continue to travel through space, a frozen planet, for an indeterminate time.

PLANETARY DAYS AND YEARS

Here on earth, we consider a day to be the time it takes the planet to rotate once on its axis—a time we usually consider to be 24 hours, but which, in fact, is approximately 23 hours, 56 minutes, 4.1 seconds. Similarly, a year is defined as the time it takes the earth to make a complete orbit about the sun—or about 365.26 days. The extra 0.26 is what requires us to have a leap year about every 4 years. (Not always *every* 4 years, as is commonly thought. The first leap year after the leap year of 1896, for example, was 1904, a span of 8 years.)

The other eight planets do not take the same amount of time as the earth to rotate around their axes or to orbit the sun. Thus their "days" are not 24 hours; their "years" not 365 days.

Planet	Planetary day (in "earth" days)	Planetary year (in "earth" years)
Mercury	58.65	0.241
Venus	246.00	0.615
Earth	1.00	1.000
Mars	0.97	1.881
Jupiter	0.41	11.820
Saturn	0.43	29.458
Uranus	0.43	84.018
Neptune	0.65	164.780
Pluto	6.30	248.400

COMETS

A comet is a celestial body of small mass and high volume that consists of a small, bright nucleus enveloped in luminous gas. A comet moves under the sun's gravitational influences. There are actually two kinds of comets: short-term comets, which complete their orbits of the sun in 200 years or less (Halley's comet, with its orbital time of about 76 years, is one of these) and long-term comets, which may take thousands of years to complete a single revolution. It is thought that all the comets were formed about the same time as the solar system itself and are among its permanent members. It is also believed that all comets were originally long-term comets. The short-term comets, then, are comets whose orbits have altered—comets, in fact, in the last stages of life. The life-span of a short-term comet is thought to be only a few million years. As comets lose material with successive passages near the sun, they fade in brightness. Some may break up entirely, their remains becoming streams of meteoroids.

METEORS AND METEORITES

The word *meteor*, taken from the Greek, means "things in the air" and, until the end of the seventeenth century, was used to include thunderstorms, clouds, and rainbows; hence, the word *meteorology*—the study of the atmosphere and its phenomena. Today, meteor is used to describe what are commonly called falling stars or shooting stars, pieces of matter that enter the earth's atmosphere from space. (Before they enter the earth's atmosphere they are called meteoroids.)

Although an estimated 80 million to 100 million meteors routinely collide with earth every 24 hours at speeds of 10 to 40 miles per second, few survive the fiery journey through the earth's atmosphere. Most last 2 to 3 seconds before they burn up and dissolve. Those that do make it through and collide are called meteorites. They sometimes cause quite a stir when they land. The largest *observed* meteorite consisted of 3,894 pounds of silicon, magnesium, iron, sulfur, calcium, nickel, and aluminum, and smashed into the earth in northeastern China in March 1976. This was just the chunk that hit, the largest of more than 100 fragments that exploded in the sky over China. It left a crater 12 feet wide and 6 feet deep. Larger meteorites have been found. The Hayden Planetarium in New York City, for example, has a specimen that weighs 34 tons. Amazingly, despite this constant bombardment, few people have been hit. The last certified meteorite victim in this country was Mrs. Hewlett Hodge of Sylacauga, Alabama. One day in 1954 she was napping on the couch in her living room when a small meteorite crashed through the roof and bounced off her head, dealing a glancing blow.

METEOR SHOWERS

A meteor shower, literally a rain of meteors in the atmosphere, consists of debris from comets whose orbits intersect the earth's. Some showers occur annually and can be predicted. They are named for the constellation in which their radiant (the point in the sky where they seem to originate) is located.

Name of shower	Annual date of maximum activity	Duration (days)
Quadrantid	January 3	1
Lyrid	April 22	1
Eta Aquarid	May 5	2
Delta Aquarid	July 29	15
Perseid	August 12	5
Draconid	October 10	0.25
Orionid	October 21	2
Taurid	November 15	30
Leonid	November 17	0.25–2
Geminid	December 14	4

ECLIPSES

Solar eclipses (eclipses of the sun) have a maximum life-span of 7.5 minutes. Lunar eclipses (eclipses of the moon) have a maximum life-span of 1.75 hours. Total solar eclipses can be seen from only a few locations on earth, whereas lunar eclipses are visible wherever the moon is above the horizon at the time of eclipse.

Atmospheric Phenomena

HURRICANES

What we call a hurricane is a cyclone in the Australian or Indian ocean and a typhoon around the East Indies and the China Sea. The general tag meteorologists give these destructive acts of nature is *cyclonic storms*. A hurricane is a great swirl of wind and rain measuring from 50 to 500 miles across. It has a peaceful "eye" in the center measuring about 15 miles in diameter. Born at sea, a hurricane travels with winds swirling about the eye at speeds of up to 200 miles an hour. About 30 to 100 hurricanes are born each year. They travel at the sedate rate of about 15 miles an hour; at their peaks, they can carry tons of water with them as they move. Hurricanes may last from several days to more than a week, but longevity is no measure of viciousness. In 1942, a cyclone that raged for 2 days killed 11,000 people in Bengal, India, while Hurricane Eloise, lasting 2 weeks in 1975, was responsible for only 71 deaths in the Caribbean and the northeastern United States.

TORNADOES

The most violent of all weather phenomena, tornadoes are dark, funnel-shaped clouds of furiously rotating air whose diameters range from a few feet to a mile. A tornado has a maximum life-span of about 9 hours; during this time, it can travel up to 300 miles, with wind velocities of 200 to 300 miles per hour.

LIGHTNING

A lightning stroke, the electrical discharge occurring during a thunderstorm between two clouds, or between a cloud and the earth, lasts 45 to 55 microseconds.

SNOWFLAKES

A snowflake takes about 10 minutes to form in the atmosphere. Depending on conditions, a snowflake can last as little as 10 minutes (some melt before reaching the ground) or as long as centuries. The shape of a snowflake is dependent on temperature: Stars form at about $-14°$ to $-17°$ C; ice needles at about 0 to $-7°$ C; plates and columns at about $-10°$ to $-22°$ C.

The Earth and Terrestrial Phenomena

CONTINENTS

According to the theory of continental drift, after the earth's crust cooled, its entire land mass formed one supercontinent, now called Pangaea, which lasted for about 2.3 billion years. Then, about 200 million years ago, this huge continent began breaking up into two smaller continents, now called Laurasia and Gondwanaland. The ages of these aren't clear because they themselves soon began to break up. Pieces of what once was Laurasia are now known as the continents of North America, Europe, and Asia. Gondwanaland broke down into what we know as South America, Africa, India, Australia, and Antarctica. So in a sense we still have those two continents with us as a jumbled global jigsaw puzzle. Just recently, geologists began to suspect there may have been an eighth continent, called Pacifica, which broke up completely about 65 million years ago, after it split away from Pangaea about 225 million years ago. According to this theory, pieces of Pacifica skidded around the globe crashing into coastlines, wrinkling the earth's crust, and creating mountain ranges, such as the Sierra and the Rocky Mountains. This theory was put forth by Amos Nur of Stanford University and Zvi Ben Avraham of the Weizmann Institute in Israel. If it is true, then Pacifica is the first truly dead continent, gone at an age of about 160 million years.

EARTHQUAKES

An earthquake is a series of elastic waves, usually caused by a rupture of the earth along a line of weakness known as a fault. Most earthquakes last only a few seconds, but some, like the great Lisbon earthquake of 1975, have lasted more than 5 minutes. What makes an earthquake so disastrous is not how long it lasts but where it hits and how high a number it gets on the Richter scale, a kind of Nielsen rating for quakes. Usually anything rated over 5 is considered pretty serious. For example, an earthquake rated 5.8 on the Richter scale killed 12,000 people in Morocco in 1960; another tremor rated 5.6 killed 500 people in Indonesia in 1976. The quake that razed San Francisco in 1906, rated a stunning 8.3 on the Richter scale, lasted for only about 40 seconds, while the famous Easter earthquake that struck Alaska in March 1964, rated at a stronger 8.5, lasted 4 minutes, but killed only 114 people, thanks to the state's low population density.

The main shock of an earthquake is often accompanied by what are called foreshocks and aftershocks. Foreshocks are small quakes that occur a few hours or days before the main quake. Aftershocks follow the main quake and may continue for as long as a year.

DESERTS

It is not only man who turns deserts into oases. Nature, if left alone, will eventually do the job. Most of the deserts of the world are confined to two belts, one at about 30 degrees north latitude, the other at about 30 degrees south. Because of the way moisture is circulated in the atmosphere, the land beneath these belts tends to be arid. Many scientists feel that the atmospheric conditions above these belts always remain the same, but that the deserts move. The theory is that the continental plates on which the deserts ride are continually shifting. The movement pushes deserts out from under these arid belts and pulls fertile, wet areas to them. The Sahara, for example, was covered by a glacier some 450 million years ago.

MOUNTAINS

Mountains take longer to form than to erode. The life cycle of a chain begins with the deposition of huge amounts of sediment along fault lines. Then, through massive shifts in the continental plates, the earth becomes deformed. Jagged edges of rock are ground together and thrust toward the sky, and mountain ranges are formed. The entire process can take anywhere from 100 million to 300 million years.

Once a mountain exists, it immediately starts to erode. Wind, rain, cold—all types of weather—take their toll, as do the continuing shifts in the earth's surface. If the mountains are small enough, moss, lichen, and then larger plants begin to grow on their rocky slopes, breaking apart the boulders and reducing them to softer, less durable dirt. After between 5 million and 50 million years, a once-proud range can be reduced to sea level again.

ROCKS

Rocks are among the few solid things that are indeed older than the hills. The oldest known rocks are believed to be about 3.7 billion years old, give or take 70 million.

OCEANS

In geological terms, oceans are remarkably young. Formed from a constant shifting of continental plates, they grow and fade at relatively constant rates. The Pacific Ocean is the oldest ocean around; the North Atlantic, which has been expanding at the rate of about 2 inches per year, is next. The South Atlantic, one of the newest oceans, is only 150 million years old; it has taken the place of even older, dead oceans (250 million years), whose traces can be found in the Caledonian-Hercynian-Appalachian mountain chains of northwestern Europe and eastern North America.

The continental plates, of course, continue to move, shrinking some oceans, creating others. Two of the youngest embryonic oceans appear to be the Red Sea and the Gulf of Aden—between the Arabian peninsula and Africa—both of which have been in existence for no more than 20 million years.

ICEBERGS

The life-span of an iceberg depends on whether it was calved in the Arctic or Antarctic. Arctic icebergs, which resemble jagged floating mountains, are calved along the coast of Greenland. They then drift on a 3-year journey to the Grand Banks of Newfoundland, where they confront the warm waters of the Gulf Stream. Here even the largest icebergs melt within 3 weeks. Antarctic icebergs are calved from the giant ice shelves that rim the Antarctic continent. These are flat, tabular bergs, or ice islands, that can be several hundred feet high and over 100 miles in length. Because of their size and the influence of the antarctic circumpolar current, the life-spans of these icebergs are indeterminate.

An iceberg's longevity is prolonged by two factors: (1) being white, it reflects 90 percent of the sunlight that strikes it, and (2) close to 90 percent of the average iceberg remains submerged in water that is close to the freezing point.

There have been several well-publicized schemes to capture icebergs and tow them to water-poor areas of the world. The most ambitious, put forth by Iceberg Transport International, Ltd., and headed by Prince Mohammed Al-Faisal, a nephew of the king of Saudi Arabia, is to capture a mile-long iceberg in the Antarctic and tow it to an offshore site in Saudi Arabia. The trip is expected to take 9 months. It would cost about $100 million and would provide a billion-ton source of drinking water that could also be used as a floating refrigerator for food storage during the estimated 3 years it would take the giant ice cube to melt.

LAKES

Lakes may "live" only a few days—consider one formed by a beaver dam—or they may last millions of years. Lakes start to die as soon as they are born. A youthful lake is oligotrophic (oxygen-rich), but it immediately begins to lose its rich concentration of oxygen and gradually becomes eutrophic. When this happens, bottom plants begin springing up and vegetation on the shore starts crowding in, slowly choking the lake until it becomes a marsh, then a small pond, and eventually, a patch of forest. Because this slow death by suffocation varies with local geography and weather, as well as with the size of the lake (scientists do not agree when a "lake" becomes a "pond"), general life-spans are hard to determine. So far, the most dependable life-spans apply to the lakes that continually form and disappear along the Lop Nor River in Central Asia. These lakes have life cycles of 1,500 years. At the end of each cycle, the river drains one lake basin and fills another.

GEYSERS

A geyser is a thermal spring that intermittently releases a column of water and/or steam that gushes skyward. Although more than 200 geysers are found in Yellowstone National Park, Wyoming, only two other regions have significant numbers of geysers—Iceland, with about 30, and New Zealand, with fewer still.

The life-spans of geysers are extremely varied. Yellowstone's Old Faithful has been reported active for about 150 years, but may, of course, have been functioning long before it was discovered. On the other hand, Excelsior Geyser, which may have been the largest in Yellowstone's history, was active only in the 1890s. One Icelandic geyser, born dramatically in 1896, survived a mere 10 days.

Traditionally, Old Faithful has erupted every 65 minutes, rarely missing its mark by more than 20 minutes. The eruption lasts for about 5 minutes, during which time the water is discharged to a height of 100 to 150 feet. In recent years, however, Old Faithful has become somewhat less regular. For some other geysers, such irregularity has signaled that the end of their eruptions was only a few years off.

In the case of one geyser, however, the natural end of its discharges was not necessarily final. Iceland's most famous geyser is the Geysir, from which all others derive their names. Located about 30 miles east of Reykjavik, the Geysir has been active for several thousand years. Though it has survived many earthquakes and volcanic eruptions that have destroyed neighboring geysers, the Geysir spontaneously went dormant in 1935. Unwilling to lose the famed monument, the Icelanders cut a lip in the geyser's basin, lowering the water table about 3 feet. The discharges began again, and the Geysir remains active (the notch has since been patched). But the eruptions are irregular now, with 2 or 3 minutes of water flow at roughly 10-minute intervals.

RADIOACTIVE ISOTOPES

Certain atoms, the so-called radioactive isotopes, are continually decaying or disintegrating into other atoms. Uranium 238, for example, decays into thorium 234, which decays into protactinium, and so on. The life-spans of these atoms are expressed in terms of their half-lives—the time it takes for half of the atoms of a particular isotope to decay. Plutonium 239, the isotope used in both nuclear reactors and atomic weapons, has a half-life of 24,400 years. In other words, if you have 1 pound of plutonium and wait 24,400 years, you will have 0.5 pound of plutonium left. The life-spans of radioactive isotopes range from less than a millionth of a second to more than trillions of years.

HALF-LIVES OF RADIOACTIVE ISOTOPES

Carbon 11	20 minutes
Carbon 14	5,580 years
Chromium 49	42 minutes
Indium 115	60×10^{14} years
Iodine 123	13 hours
Iodine 136	8.6 seconds
Krypton 40	1.2×10^{9} years

Lanthanum 138	2.0×10^{11} years
Lutetium 176	7.5×10^{10} years
Neodymium 144	1.0×10^{15} years
Nitrogen 13	10.1 minutes
Polonium 212	3×10^7 seconds
Rhenium 187	4.0×10^{12} years
Rubidium 87	6.0×10^{10} years
Samarium 147	1.4×10^{11} years
Technetium 99m	6.0 hours
Tellurium 130	1.0×10^{21} years
Thorium 232	1.4×10^{10} years
Uranium 235	7.1×10^8 years
Uranium 238	4.5×10^9 years
Uranium 239	23.5 minutes

ENERGY RESOURCES

The world has grudgingly admitted that the energy crisis is not merely a passing fad, but a cruel reality. Chastened, we are taking stock of what's left. Estimating how much time we have until present deposits of oil, coal, uranium, and so forth are depleted is an especially tricky business. For one thing, no one really knows exactly how much is left. For another, the world's energy appetite in the years to come may either exceed or fall short of current forecasts. Lastly, changing energy trends—relying more on nuclear power than on coal, for example—may also affect life-spans of individual resources.

To get some idea of the relative longevity of known as well as estimated resources, physicist Dr. Ted Taylor and coauthor Charles Humpstone, in *The Restoration of the Earth*, worked out a supply chart based on two conditional factors: the estimated energy appetite of the world in the year 2000, and the assumption (for the sake of simplifying the estimate) that each resource will be the *only* source of energy available to satisfy that appetite. Using a standard energy yardstick—metric tons of coal equivalent (MTCE)—the two authors estimated that by the year 2000 the world's energy appetite will have increased from the 7.5-billion-ton equivalent it consumed in 1970 to 32 billion MTCE. Based on that gluttonous projection, the accompanying longevity table of known and estimated deposits was computed.

POTENTIAL NONRENEWABLE SOURCES OF ENERGY REMAINING BY THE YEAR 2000

| | *Remaining number of years* | |
Source	*From known concentrated deposits*	*From estimated reserves at low concentration or undiscovered concentrated deposits*
Coal	26.0	500
Petroleum	2.5	34
Shale oil	1.0	110
Natural gas	3	35
Uranium	100	6,000,000
Thorium	70	10,000,000
Lithium 6	100	10,000,000
Deuterium (from the sea)	10,000,000,000	10,000,000,000

PLANTS AND FLOWERS

In the animal world, defining an organism's life-span could hardly be easier: It is simply the time from birth to death. Both events are easily identified, and figuring the period between is mere subtraction. If you watch enough individuals, you can calculate an average life-span for the species and make at least a reasonable estimate of the maximum possible longevity.

Among plants, the problem is far more complex. Take a simple alga, for example. Algae sometimes reproduce sexually; male and female gametes, each carrying half the genetic message needed for a whole organism, unite and form a new individual, a process just like reproduction among animals. In this case, the "birth" of the daughter cells can be pinpointed, and the mother cell's span is still calculated from its time of birth. But algae, and many higher plants, also reproduce asexually. A mature alga can divide into two or more daughter cells, each identical to the parent. At this point, the concept of an individual life-span begins to break down. The mother cell has not exactly died, but by convention the age of both daughters is set at zero when division occurs. At best, this is a somewhat arbitrary convenience. When a rabbit gives birth asexually, as has occurred in some laboratory experiments, the mother remains herself; the daughter is clearly a new creature. Whatever inner clock governs the mother's longevity still runs down pretty much as it would have had the birth never taken place. But algae die when eaten or when their water dries up or becomes polluted—not on a fixed schedule. As algae lives are tolled, from division to division, the organisms survive only a few hours or days. But in a very real sense, through their daughter cells, algae are immortal.

For higher plants, life-spans seem easier to pin down, at first. In climates where the seasons vary markedly, tree trunks and the stems of woody shrubs show clear rings where the large, thin-walled wood cells produced in spring alternate with dark bands of the small, thick-walled cells that grow in harsher seasons. Where the contrast is clear, the pairs of alternating bands can be counted to give a fairly accurate idea of how long the tree has been growing.

Smaller plants are categorized as annuals, which live through only one growing season; biennials, which grow vegetatively in one season and flower and die in the next; and perennials, which live for several years and usually bloom repeatedly. Most flowers and ornamental plants are either annuals or perennials. A number of farm crops are biennials, though most are picked at the end of their first season and are raised as though they were annuals. Soft-tissued plants lack growth rings, so there is no obvious way to tell the age of an herbaceous perennial.

This simple scheme breaks down in some cases. The castor bean, for example, grows as an annual in temperate climates; it dies with the autumn frost. In the tropics, however, castor beans are woody perennials. Then there is a colony of box huckleberries in Pennsylvania. Well over 1 mile across, the colony is thought to be 13,000 years old. Though each plant appears to be an individual, they are all linked, each growing from the same spreading root structure. Somewhere in the colony, the original stem may still survive. If not, can it truly be said to have died while so many of its offshoots live on?

Interesting as the philosophical question is, most plants have well-recognized pragmatic life-spans. But, except for trees, botanists have not compiled precise life-span data on plants. Dividing nontree plants into annuals, biennials, and perennials apparently is sufficient for their needs. What farmer cares that a crop plant could theoretically survive for decades when he plans to harvest it after a season or two or when, like asparagus, it becomes woody and inedible after a few years? Do you really care that your flower garden might still be blooming when the mortgage is finally paid off if the garden is standing where you plan to put a patio next summer? More useful is the concept of vigor, a subjective catch-all that totals all those factors that contribute to a plant's health and survival. What vigor denotes is not so much how long the plant will survive as an idea of how sturdy it is, how well it is doing at the moment, and how well it will endure stress. The idea applies to varieties as well as to individual plants. Will this strain of melon compete successfully against encroaching weeds, tolerate a dry spell, and survive insect attack? If so, the strain is vigorous. But if your philodendron's leaves are yellow and wilting and it hasn't sent out a new runner in a year, it is losing its vigor and may not long survive unless you act promptly to cure whatever ails it.

For annual plants, the life process is self-limiting. If the soil is fertile and light levels, humidity, and water supply are correct, if the garden or flower bed is weeded and insects are kept down, the annual will likely grow and flower on schedule. At the end of the season, it will die, no matter how much care it receives.

Perennials, however, often require more complex care if they are to reach their maximum life-spans. This is especially true of plants raised indoors, where light levels vary from one part of a room to another, humidity is usually less than ideal for plants, and dust and dirt that can clog leaves' respiratory pores are never washed away by the rain. Unlike most farmers, for whom economics dictates that plant care be reduced to a straightforward series of simple steps geared to mass production, many amateur gardeners find that keeping their perennials in top condition becomes a never-ending bustle from plant to plant, chore to chore. We'd like to help, but it would take a small encyclopedia to solve even the most basic problems any household horticulturist is likely to face. What we can do is tell you which common flowers and vegetables are perennials and which are shorter-lived. Most of the common ones are listed in the accompanying charts.

Crop plant	Life-span	Crop plant	Life-span
Asparagus	Perennial	Parsnips	Biennial
Beans	Annual	Peas	Annual
Beets	Biennial	Potatoes	Perennial
Buckwheat	Annual	Rhubarb	Perennial
Cabbages	Biennial	Soybeans	Annual
Carrots	Biennial	Strawberries	Perennial
Celery	Biennial	Tobacco	Annual
Corn	Annual	Tomatoes	Annual
Flax	Annual	Wheat	Annual
Jute	Annual		

Flower	Life-span	Flower	Life-span
Alyssum	Perennial	Lupines	Perennial
Canterbury bells	Biennial	Nasturtiums	Annual
Chrysanthemums	Perennial	Peonies	Perennial
Cosmos	Annual	Petunias	Annual
Dahlias	Perennial	Phlox	Perennial
Delphiniums	Perennial	Sunflowers	Annual
Foxgloves	Biennial	Sweet peas	Annual
Iris	Perennial	Tulips	Perennial
Larkspurs	Annual	Wallflowers	Biennial
Lavender	Perennial	Zinnias	Annual
Lily of the valley	Perennial		

TREES

Non-Fruit-Bearing Trees. These trees generally live for a long time if they are not cut down or destroyed by some disease or natural disaster. Nonetheless, there is a tremendous range between, say, the life-span of a poplar and that of a fir.

Non-fruit-bearing tree	Life-span (years)	Non-fruit-bearing tree	Life-span (years)
Southern poplar	100	Longleaf pine	225
Red maple	110	American elm	235
Northern catalpa	115	Shortleaf pine	235
Dogwood	115	White fir	260
American linden	120	Shore pine	275
American holly	125	Sugar maple	275
Loblolly pine	175	Cascades fir	275
Slash pine	200	American beech	275
Pacific yew	200	White ash	275
Eastern black walnut	200	Red spruce	275
Arizona cypress	200	Western larch	325
Grand fir	225	California cedar	350

Non-fruit-bearing tree	Life-span (years)	Non-fruit-bearing tree	Life-span (years)
Eastern hemlock	350	Bald cypress	600
Red fir	350	Douglas fir	750
Noble fir	400	Redwood	1,000
Sugar pine	400	Giant sequoia	2,500
Ponderosa pine	450	Bristlecone pine	3,000+
White oak	450		

Fruit Trees. Considering the contribution they make to sustaining the planet's population, fruit trees, particularly noncitrus ones, deserve longer life-spans than those dealt out to them.

The information in the accompanying chart is based on the assumption that the trees are in excellent condition and are receiving the best of care. The date on which the tree begins to bear fruit depends upon the sapling's age when transplanted; the life-spans given assume the transplanting of young (3- to 4-year-old) saplings.

Fruit tree	Fruit-bearing age (years after transplanting)	Life-span (years)
Apple		
Dwarf	2–3 (when tree is 5–7 years old)	20–25
Semidwarf	3–4	25+
Standard	3–5	25+
Apricot		
Semidwarf	2–3	12–15
Standard	2–3	12–15+
Cherry		
Sweet	2–3	25+
Standard sour	2–3	20+

Fruit tree	Fruit-bearing age (years after transplanting)	Life-span (years)
Semidwarf Montmorency	2–3	20+
Dwarf Northstar	2–3	20+
Grape vine	2–4	20
Nectarine		
Dwarf	2–3	8–12
Standard	2–3	12+
Peach		
Minipeach	2	8–10
Dwarf	2–3	10–12
Standard	2–3	12+
Pear		
Dwarf	3–4	25+
Standard	4–5	25+
Plum		
Semidwarf	3–4	12+
Standard	3–4	10–12

Citrus trees start bearing fruit at about 4 years of age and continue to live and bear fruit (under ideal conditions) for as long as 100 years.

Bonsai Trees. Bonsai trees, the plant miniatures that are products of a uniquely Japanese art form, are both natural and man-made phenomena. In fact, the idea of taking a cutting from a tree, keeping it well pruned, confining its roots, and sculpting around it a tiny, three-dimensional representation of its natural setting emerged from the discovery that such miniatures did exist in nature—that tiny, perfectly formed trees had actually found ways to carve minuscule natural niches in the hollows of otherwise inhospitable rock formations.

Bonsai trees live as long as their giant counterparts, if they are well cared for. Artificial bonsai trees two and three centuries old are not uncommon; natural bonsai around 1,000 years old have been found, transplanted, and nurtured.

CUT FLOWERS

Cut flowers can be preserved in full bloom for as long as 2 weeks (depending on the variety) if they are well cared for. Keeping them beautiful requires minimal effort: recutting (on the bias) daily; fresh water every day; some chemical nutrients, found in any florist's shop; and a relatively cool spot to sit, away from direct sunlight. If these conditions are provided, fresh-cut roses last an average of 7 to 10 days; carnations, 2 weeks; gladiolas, 7 to 10 days; and pompoms, 2 weeks. Flowers purchased from a florist last an average of 2 to 3 days less, because it took that long for them to be shipped to the shop.

Cut flowers kept out of water will die very quickly, but if they are thoroughly watered and refrigerated just before they are used, they will wilt more slowly. The reason: They go into shock if they are used to a plentiful supply of water that is suddenly cut off. It then takes a while before the processes of degeneration catch up with the trauma of a waterless existence.

4 / FOODS AND BEVERAGES

IN 1809, Napoleon awarded a prize of 12,000 francs to a French confectioner named François Appert for his invention of a technique that would extend the life-spans of various foods. The technique was essentially canning (Appert, who also invented the bouillon cube, employed glass jars rather than cans), and it has been said that if Napoleon had been able to confine the innovation to France he would have conquered the world. Within a year of Appert's breakthrough, however, Bryan Donkin was granted an English patent for his "tin can," leading the way for Jacob Perkins's refrigerator in 1834, Gail Borden's condensed milk in 1858, and Clarence Birdseye's development of quick freezing in the 1920s.

All foods begin to deteriorate at the moment of harvest or slaughter. In some, a certain degree of deterioration is desirable, thus we speak of "aging" such things as meat, cheese, and wine. In others, notably vegetables, fresher is better.

Food deteriorates or spoils for two reasons: chemical changes within the food itself, and the growth of outside organisms, specifically molds, yeasts, and bacteria. Although chemical changes within food may cause wilting, discoloration, softening, loss of flavor and nutritional value, they are not in themselves dangerous. Nor are yeasts. Only bacteria and molds can directly result in food poisoning or illnesses and possibly death.

Bacteria cannot grow in strong acids, so such tart foods as fruits and pickled products present no danger. Low-acid foods—meats, poultry, dairy products, and vegetables—however, provide ideal environments for bacteria and should be treated with caution. The only way to destroy such bacteria is to heat the food to a sufficiently high temperature (over 190° F). Food in liquid form (or food to which water can be added) should be brought to a boil and then simmered for at least 10 minutes. Solid foods—chops, for example—should be heated in the oven or on top of the stove until the surface sizzles and chars. Remember that water at 212° F, the boiling point, transfers heat more efficiently than an oven at 212° F. (If you doubt this, stick your hand in a 212° F oven and then in a pan of boiling water.) So make sure suspect foods are *thoroughly* cooked in the oven.

We can prolong the life-spans of the foods we eat in a number of ways, all designed to prevent or arrest the growth of organisms that cause spoilage. We dry food because organisms require moisture to multiply; we salt-cure and sugar-cure food because bacteria and molds do not grow in high concentrations in either substance; we smoke food because the traces of formaldehyde present in wood smoke arrest bacterial development and we pickle food because the acidic environment will not sustain bacterial growth.

Food properly canned, either commercially or at home, will not spoil during storage if the can's seal is unbroken; according to industry sources the food will last upwards of 50 years. But while no dangerous changes will occur in intact canned food, there will be distinct changes in quality (taste, texture, color, food value) over time. Most canned goods can be stored up to 1 year at room temperature with only slight changes in quality. For longer periods they should be stored at 45° to 50° F. While open cans and jars may be unattractive, no harm can come from using them for temporary storage in the refrigerator.

Since World War II, the primary home device for extending the life-spans of foods has been, of course, the refrigerator/ freezer. Freezing is not only the quickest, easiest, and cheapest way of preserving food, but is also the process that least alters flavor, color, and nutritional value. Most refrigerators provide temperatures of about 40° to 50° F for foods to be kept cold but unfrozen, 25° to –10° F for frozen foods. The ice-cube compartment often found in older refrigerators will hold frozen foods for a short time, but since its temperature is seldom below 20° F it is not recommended for extended storage. A true freezer, separate from the large storage area of the refrigerator, should provide a temperature of 0° F or below.

The temperature of frozen food is critical. As pointed out by the Department of Agriculture, string beans can be stored 1 week at 30° F, 1 month at 20° F, 6 months at 10° F, 1 to 3 years at 0° F, indefinitely at lower temperatures. (It should be noted that self-defrosting or frost-free refrigerators work by taking moisture from the air they hold, causing foods to dry out and sometimes shortening their life-spans.)

Never refreeze food that has thawed to 40° F internal temperature. Partially thawed food registering below 40° F—or still showing ice crystals—may safely be refrozen.

The life-spans given in this chapter refer to the maximum recommended period during which the food can be safely stored without a significant loss of quality. The temperatures involved in the various storage procedures are as follows:

Room temperature	70° F
Refrigerator	50° F
Ice-cube compartment	25° F
Freezer	0° F

Freeze-dried Food

Freeze-drying is a process first patented in 1934 that removes all but 2 percent of the water in a particular food; hence, the main enemy of freeze-dried food is moisture. As a freeze-dried product absorbs moisture from the air, it begins to turn brown and deteriorate in quality. In the case of powdered freeze-dried products like coffee or fruit juices, the food cakes and becomes so rocklike that it simply cannot be rehydrated. Powdered freeze-dried foods often have packed with them little cardboard cylinders marked "Do Not Eat." The cardboard contains a desiccant that will keep the food from caking for a little longer.

Freeze-dried foods stored at low temperatures keep longer than those stored at high temperatures. They also keep better in cans because ordinary glass or plastic jars with screw tops cannot keep moisture out efficiently. Once the seal on a container of freeze-dried food is broken, the food's shelf life depends upon the conditions under which it is stored. The food will deteriorate faster on the hot, humid shelf over the stove than on a pantry shelf and will keep longest in the refrigerator.

Freeze-dried soups, sauces, and entrées containing more than one ingredient have shelf lives equal to the average of the shelf lives of their ingredients. Because foods are dehydrated to different degrees, the drier ingredients tend to absorb moisture from the wetter ones. (The carrots in a freeze-dried beef-stew dinner may lose moisture to the neighboring bits of beef, slightly extending the "maximum quality shelf life" of the carrots and slightly shortening that of the beef.)

So far, the main consumers of freeze-dried foods seem to be the Mormons (who stock their cellars with it), the military, and campers. Some of the foods used by the army, for example, are packed with nitrogen gas rather than air. It seems that food molecules that are freeze-dried tend to have points of weakness where oxygen can attack and spoil the food. Packing the food in nitrogen loads the weak points with nitrogen molecules, leaving no room for the oxygen.

Freeze-dried foods are expensive, but a little goes a long way. For example, 1 pound of freeze-dried beans can produce 12 to 13 pounds of rehydrated vegetables; 4 ounces of freeze-dried beef produces 1 pound of beefsteak when rehydrated; and 6.5 ounces of freeze-dried chicken stew yields 2 pounds.

SHELF LIVES OF FREEZE-DRIED FOODS

Food	Shelf life (months)		
	At 40°F	At 70°F	At 90°F
Steak	72	36	18
Chicken	72	36	18
Fish squares	72	36	18
Scrambled-egg mix	60	36	18
Apples and apple juice	72	36	18
Carrots	36	18	9
Cottage cheese	24	12	6
Cherries	48	24	12
Chives	24	12	6
Corn	48	12	6
Potato granules	72	36	18
Potato slices	36	18	9
Onions	48	24	12

Meat

FRESH MEAT

There are definite ways of caring for and storing meat that will extend its life-span, as well as protect its taste and nutritional value. First, choose your butcher wisely; then allow as little time as possible to elapse in transit between the butcher shop and the freezer.

Before storage, all fresh beef, lamb, veal, and pork should be wiped dry with a damp cloth or paper towel and rewrapped. They should then be placed in the coldest part of the refrigerator or in the freezer. In general, the larger the cut of meat, the longer its life-span.

Meat kept unfrozen in the refrigerator should be loosely wrapped to allow it to "breathe" (air circulation keeps the outside of the meat dry and inhibits bacterial growth). Some wrapping papers available are oxygen-impregnated and release their oxygen to the meat, increasing its life-span; plastic wrapping inhibits the partial surface drying that extends the meat's life-span and should, therefore, be avoided.

Meat stored in the freezer, however, should be wrapped in airtight, moisture-proof materials such as aluminum foil, a heavy-duty plastic bag or wrap, or specially coated freezer paper. Ordinary waxed paper and butcher's paper are not adequate.

To store after cooking, meat should be cooled rapidly, placed in an airtight, moisture-proof wrap or container, and placed in the refrigerator or freezer. It should not be stored before it has cooled to room temperature, since condensation will form inside the wrapping, spawning bacterial growth; moreover, if it is still hot, the meat will raise the temperature of the refrigerator itself, decreasing the life-spans of other foods.

The accompanying tables indicate the life-spans for different types of meats stored under a variety of conditions.

STORAGE LIFE-SPANS FOR BEEF AND LAMB

	Life-span		
Type of beef or lamb	On refrigerator shelf (days)	In ice-cube compartment (weeks)	In freezer (months)
Lamb chops	2–4	2–3	6–7
Beef steaks	2–4	2–3	6–8
Roasts	3–6	2–3	8–12
Ground	1–2	1	3–4
Cut up (for stews)	1–2	2–3	6–7
Lamb bones	1–2	2–3	6–7
Bone marrow	1–2	1	1
Cooked (leftovers)	4–5	8–12	3–6

STORAGE LIFE-SPANS FOR PORK

	Life-span		
Type of pork	On refrigerator shelf (days)	In ice-cube compartment (weeks)	In freezer (months)
Chops, cutlets, spareribs	3	2–3	2–3
Roasts	4–6	2–3	4–8
Ground and variety meats	1–2	2–3	1–3
Cooked (leftovers)	2–3	2–3	2–3
Cured whole ham	14	Don't freeze	Don't freeze
Cured half ham or end	7	Don't freeze	Don't freeze
Cured sliced ham (country style)	3	Don't freeze	Don't freeze

STORAGE LIFE-SPANS FOR VEAL

	Life-span		
Type of veal	On refrigerator shelf (days)	In ice-cube compartment (weeks)	In freezer (months)
Chops, steaks, cutlets	2–3	1–2	4–8
Roasts	3–5	1–2	6–7
Ground or cut up (for stews)	1–2	1–2	2–3
Cooked (leftovers)	4–5	1–2	2–3

STORAGE LIFE-SPANS FOR LUNCHEON MEATS

Type of luncheon meat	Life-span			Special instructions for storage
	On refrigerator shelf (days)	In ice-cube compartment (weeks)	In freezer (months)	
Bacon	5–7	2	Slab: 1–3	Keep tightly wrapped.
Bologna	Sliced: 3–5 Unsliced: 5–7	8–12	Don't freeze	Refrigerate in original package or wrapped loosely in waxed paper.
Frankfurters	3–4	Don't freeze*	Don't freeze*	
Knockwurst	7	Don't freeze*	Don't freeze*	
Liverwurst or liver sausage	7	Don't freeze*	Don't freeze*	
Salami	Sliced: 2–3 Dry, with casing cut: 14	Don't freeze*	Don't freeze*	When sliced, be sure slices are flat before storage; wrap in moisture-proof, vaporproof paper. Unsliced, dry or semi-dry salami, with casing uncut, can be stored indefinitely at room temperature.
Sausages	Fresh: 2–3 Cooked or smoked: 4–6	8–12	Don't freeze*	
Sausages (dry)	With casing cut: 14	Don't freeze*	Don't freeze*	Refrigerate once casing is cut. With casing uncut, can be stored indefinitely at room temperature.

*Because of their high fat content, many luncheon meats and cold cuts do not hold up well to freezing.

STORAGE LIFE-SPANS FOR VARIETY MEATS

		Life-span			
Type of variety meat	Raw, on refrigerator shelf (days)	Cooked and covered, on refrigerator shelf (days)	In ice-cube compartment (weeks)	In freezer (months)	Special instructions for storage
Brains	1	1	2–3	3–4	Must be firm and of good color when purchased; blanch before freezing.
Kidneys	1	2–3	2–3	6	
Liver	1–2	2–3	2–3	6	Store loosely wrapped.
Pancreas (sweetbreads)	1	1	2–3	3–4	Use immediately, if not freezing; blanch before freezing.
Spareribs	3	2–3	2–3	3–4	Store unwrapped or loosely wrapped in coldest part of refrigerator.
Tongue	Fresh:1 Smoked, uncooked: 3	Fresh: 2–3 Cured or pickled: 7	Don't freeze	Don't freeze	Avoid freezing because of high fat content.
Tripe	Fresh: 1 Pickled: 2–3	1	2–3	3–4	Cook before freezing.

119

CANNED MEAT

Canned meat has a long life-span if the can remains unopened. Some canned meats require refrigeration; others do not. The best idea is to read labels carefully and observe expiration dates.

Once opened, canned meat can last 1 to 2 days at room temperature, 4 to 5 days in the refrigerator, 2 to 3 weeks in the ice-cube compartment, 6 to 12 months in the freezer. Contrary to popular belief, canned meat can be safely stored in the original can if it is resealed with an airtight wrap or cover.

PREFROZEN MEAT

If allowed to thaw, meat purchased already frozen should be eaten as soon as possible. Refreezing, while inadvisable, is not dangerous as long as the meat has not already spoiled. However, it results in a dry, distinctly inferior quality of meat. Prefrozen meats, stored in the home freezer, have the following approximate life-spans.

Type of meat	Life-span (months)
Ground beef and chopped steak	4
Beef roasts and thick steaks	12
Ground lamb	4
Lamb roasts	9
Pork chops	4
Pork roasts	8
Pork sausage	2
Cured pork	2
Veal cutlets and chops	9
Veal roasts	9
Cooked meat (prefrozen meat pies, TV dinners, and Swiss steak)	3

GAME

Moose, venison, elk, and the like can be kept in the freezer for 4 to 6 months. The meat should first be hung in a cooling refrigerator (temperature 30° to 35° F) for 10 days to 2 weeks, so that it "cures" properly before it is cut up for the freezer. Rabbit keeps 2 days in the refrigerator, 1 week in the ice-cube compartment, 2 to 3 months in the freezer, and 1 to 2 days once cooked.

Poultry

FRESH POULTRY

Fresh chicken, turkey, duck, and other edible fowl should be wiped dry with a damp cloth or paper towel and refrigerated or frozen in fresh, moisture-proof wrappings. Uncooked stuffed birds should not be kept in the refrigerator for more than 1 day, but they can be frozen. Precooked stuffings—those that contain giblets or sausage, for example—should be cooled before the bird is stuffed.

Game birds that are plucked, cleaned, and washed and dried thoroughly can be frozen for up to 4 months.

Fish

FRESH FISH

Fish and shellfish are among the most perishable foods, but with careful handling and storage their nutritional value and taste need not be impaired. In any unfrozen form—but properly iced—most fish delivered within 48 hours of being caught have shelf lives of about 5 days. Seafood, however, can take 10 days or more in transit before reaching the market. If you inspect a fish carefully before purchasing it and then store it, cleaned and loosely wrapped, only 1 or 2 days in the refrigerator, it will remain fresh and have no strong or rancid flavor. Keep in mind that a fish with a high oil content—mackerel,

STORAGE LIFE-SPANS FOR POULTRY

Type of poultry	At room temperature (hours, except as noted)	On refrigerator shelf (days)	*Life-span* In ice-cube compartment (weeks)	In freezer (months)	Special instructions for storage
Chicken	Whole, unstuffed: 24 Cut up: 5–6	1–2	1–2	12	—
Duck	Whole, unstuffed: 24 Cut up: 5–6	2–3	1–3	3–6	—
Goose	Whole, unstuffed: 24 Cut up: 5–6	2–3	2–3	6	—
Turkey	Whole, unstuffed: 24 Cut up: 5–6	Whole: 6–8 Cut up: 4–5	2–3	6	—
Giblets (gizzard, heart, liver)	1	1	1	2–3	Refrigerate immediately, loosely wrapped in waxed paper in coldest part of refrigerator.
Smoked (all kinds)	48	10–20	2–3	6	
Cooked (all kinds)	Serve as soon as possible	2–5	2–3	With gravy: 3–6 Without gravy: 1 Fried: 4	Remove stuffing and store separately; cover tightly.

Item					
Stuffed birds	4-6	1	2-3	6	—
Canned	Unopened: 1 year	Opened: 4-5	Opened: 8-12	Opened: 6	Read label carefully and note expiration date.
Prefrozen	Serve as soon as possible	5	3-4	6 Fried chicken: 4	Never store thawed poultry at room temperature.
Cooked, prefrozen (e.g., chicken and turkey dinners or pies)	—		3-4	6	—

trout, or herring, for example—turns rancid faster than a nonoily "white" fish.

Of course, the freshest fish is the one you catch yourself. Just be sure to clean, dress, and chill your catch carefully. By immediately removing the intestines, liver, heart, and gills, you eliminate the largest source of bacterial contamination. Furthermore, the digestive enzymes in many fish—particularly bluefish and other highly predacious species—are extremely powerful: They will attack and break down the walls of a fish's body cavity if not quickly removed. After the fish has been gutted, and if more than a few hours are to elapse before refrigeration, the fish should be packed in a portable ice chest.

FREEZING FISH

Not all seafood freezes well. Watery fish are particularly tricky because they tend to become mushy and flavorless. Frozen oily fish may become overly strong and gamy. Sardines and squid should never be frozen. Still, freezing fish is preferable to canning. A fresh fish, properly cleaned and tightly wrapped in an airtight package, can be frozen—best at 0° to 16° F—for 4 to 6 months for lean white fish, 3 to 4 months for oily fish. The fish should be thawed in the refrigerator or in a basin of cool water and then used within 24 hours.

There is another freezing method that can extend the life-span of frozen fish. Oily saltwater fish—and shellfish—if frozen in airtight containers filled with a 2.5 percent brine solution and then sealed, can be stored for up to 9 months. The ice that forms around the fish protects it from air, and the seal prevents moisture loss.

Commercial packers generally can freeze fish more successfully. The best method is flash freezing, either aboard the fishing vessel itself or in a large commercial plant. The process uses liquid nitrogen; at a temperature of –290° F, the fish is frozen within 5 to 8 minutes. All oxygen is excluded, and moisture loss is kept to a minimum. Prefrozen fish should be stored at temperatures below 10° F.

CANNED FISH

Not all fish are suitable for canning, but certain kinds fare rather well. Anchovies, mackerel, salmon, sardines, tuna, lobsters, crabs, caviar, eels, and herrings are the most likely candidates to leave the wide-open seas for the narrow confines of a vacuum-packed can. Canned fish will last up to 1 year unopened. Once opened, canned fish can be stored for 3 to 4 months in the freezer, 3 to 4 days in the refrigerator, or 5 to 6 hours at room temperature.

PASTEURIZED FISH

If you who don't want to sacrifice too much of the flavor and quality of fresh fish but nonetheless appreciate the availability and convenience provided by cans, you might do well to choose pasteurized fish, especially if you like crabmeat, oysters, scallops, or caviar. A pasteurized fish is hermetically packed in a can or glass jar and preserved either by heating at temperatures below 212° F or by radiation from radioactive isotopes or an electron source, processes that destroy most bacteria. Pasteurized fish, unlike ordinary canned fish, must be refrigerated (but never frozen). The process extends shelf life while retaining the desirable texture of fresh fish.

SMOKED FISH

Although the resulting qualities of texture and flavor vary considerably, any species of fish can be smoked, by either the hot-smoking or cold-smoking method. Smoking both firms and dehydrates the fish, and the salt or brine used in the process acts as a preservative inhibiting bacterial growth. Hot smoking, at temperatures ranging from 120° to 180° F for 6 to 12 hours, extends the shelf life only slightly; the fish will keep just a few days unless refrigerated, frozen, or canned. Cold-smoked fish is not cooked, but is cured through drying at temperatures ranging from 70° to 90° F for an extended period of time—anywhere from 36 hours to 3 weeks. How long the fish can be preserved depends on the strength of the brine or the amount of salt used in the process, as well as on the duration of the smoking period. In general, heavily smoked fish can be kept at room temperature for 3 to 4 weeks, in the refrigerator for 2 to 3 months, or in the freezer for up to 1 year. An important exception: Smoked mackerel is nearly as perishable as fresh fish; if it is not to be used relatively soon, it should be frozen at temperatures below 0° F—and can then last for up to 6 months.

PICKLED FISH

Although the term *pickled fish* is often applied to fish cured in brine, strictly speaking, pickling requires the use of vinegar or some other citric acid. Practically all fish can be pickled, but the process is most commonly limited to herrings. The mildest method of pickling—seviche—is not really pickling at all, but a marinating process that "cooks" the fish by the chemical action of the marinade; it doesn't appreciably extend shelf life. Another method—in which the fish is cured for no more than 2 hours, then spiced for about 3 days—extends the shelf life to about 1 week. If fish is cured for a longer period—3 to 5 days—bacterial growth is substantially retarded, and the fish will keep for up to 6 months if stored below 50° F. The stronger the acetic acid content of the pickling solution, the longer the shelf life.

Mackerel, cod, herring, trout, salmon, bass, and several other types of fish are often pickle-cured. Fillets are soaked in brine, coated with salt, packed in more salt for 2 to 10 days, then removed and scrubbed or soaked. They can then be repacked and stored for up to 9 months at low temperatures.

CAVIAR

Caviar, the roe of various sturgeons, is a delicacy with a long history, its unique taste having been celebrated since the thirteenth century. The finest caviar is made from the roe of the yellow-bellied sterlet; the "golden caviar" of the czars, it was reserved exclusively for the Russian Imperial Court, where it was served with much ceremony, accompanied by music and, no doubt, ice-cold vodka.

One of the reasons caviar is so expensive is that it takes 9 to 15 years for a female sturgeon to reach sexual maturity—20 years for the beluga, from which today's most prized berries are gathered. Another reason: Domestic sturgeon, once plentiful, are now relatively scarce after a century of greedy exploitation. No wonder a keg (135 pounds) of beluga caviar now costs about $15,000.

If you invest a small fortune in caviar, careful attention to its storage is certainly worth the effort. Caviar must never be frozen, for upon thawing the berries will burst and a container filled with mushy black sludge is all that will remain. The proper temperature for long storage is 26° F. At this temperature caviar will retain its delicacy for many months. A tin should be turned frequently to keep the coating of fat well distributed. If you expect to be serving the berries within a few weeks, they can be refrigerated at a temperature between 38° and 40° F. Pasteurized caviar—the kind that comes in tiny glass jars—should be chill-stored at 33° to 35° F and should be served before 6 months have elapsed.

Shellfish

You cannot be poisoned as a result of eating "spoiled" shellfish. Shellfish poisoning comes either from bacteria and viruses (like hepatitis) that flourish in polluted water or from the flagellate bacteria present in red tides. When shellfish are harvested from clean waters, it won't hurt to eat them after they get old; but you probably won't enjoy the experience, because shellfish carry a fantastically strong group of enzymes that makes them disintegrate—and become watery and unappealing—very rapidly.

LOBSTERS AND CRABS

Fresh lobsters should be purchased live, either from a bed of ice or a water tank (tanked lobsters contain a few ounces of water and thus are more expensive). Make sure the tail curls under when the lobster is picked; otherwise it is not fresh. Don't try to keep lobsters alive in fresh water—they will suffocate. Although a lobster should be cooked and eaten as soon as possible, it can be frozen for 4–6 months; however, the loss in quality is considerable. The best procedure is to boil the lobster for 10 to 20 minutes in a brine solution, remove the meat, pack it in brine-filled jars, and freeze at 10° F. Cooked lobster can be refrigerated for 1 to 2 days; canned, it will last 1 year.

Crab meat is similar to lobster meat in shelf life. Prefrozen Dungeness crabmeat should be used within 3 months, but prefrozen king crabmeat can last 10 months.

MUSSELS, CLAMS, AND OYSTERS

These bivalves can be kept, alive, for several days, as long as they are stored wet and under refrigeration, preferably at 40° F. Shucked and raw, they can be stored on ice for 1 to 2 days. Mussels, clams, and oysters stand up poorly to freezing, since they tend to become watery (the belly tissue disintegrates). However, shucked, purged (by soaking in a solution of cornmeal and salt water), and retained in their liquid, they can be frozen at low temperatures and stored for 3 months— though they will not be in peak condition when thawed. Surf clams, on the other hand, can be frozen more successfully because what is eaten (and frozen) is only the muscular "foot" of the clam, not the entire animal.Smoked bivalves can be frozen as well, or they can be refrigerated for several weeks in jars filled with cottonseed, peanut, or olive oil.

SCALLOPS

A scallop cannot be stored alive because it cannot hold its shell firmly closed. If unable to move through the water, it rapidly loses body moisture and dies. Scallops, therefore, should be shucked as they are caught (only the muscle is used for food). Refrigerate fresh raw scallops, loosely covered, at 32° to 35° F for no more than 2 days. Frozen scallops can be kept in the ice-cube compartment for 7 days; in the freezer they can be stored for 4 months (it's best to cook them in a court bouillon first).

SHRIMPS

Shrimps, in endless varieties and sizes, are popular the world over and are marketed in almost every conceivable way: "green" (fresh) or frozen—either in their shells or shucked and deveined—smoked, dried, canned, or vinegar-cured. Fresh shrimps can be stored in the refrigerator for 1 or 2 days; if cooked, they will keep there for 3 to 4 days. Frozen, either raw or cooked, they can be placed in the ice-cube compartment for 1 week or in the freezer for 1 month. (Do not thaw before cooking.) Canned shrimps should be used within 1 year; once opened, canned shrimps keep for 2 to 3 days in the refrigerator.

Eggs

Fresh eggs will last at least 2 days at room temperature and at least 10 days in the refrigerator—often much longer. (Fertilized eggs should not be refrigerated.) Store eggs, large end up, in the carton they come in; the little egg pockets in the refrigerator door generally do not provide enough space for the eggs to breathe. (If an egg rattles when shaken, it is not fresh; if it sinks when placed in water, it is.) Eggs should be kept away from strong-smelling foods because they absorb odors easily.

Absolutely fresh eggs can be frozen for up to 1 month in the ice-cube compartment or for 1 year in the freezer—provided they are cracked open and the yolks and whites mixed. Egg yolks *or* whites can be kept 6 to 10 hours at room temperature or up to 2 days in the refrigerator. It is best to cover the yolks with a little water in a closed container and drain off the water before use. Yolks can be frozen, although not as successfully as whites, and kept for 3 to 4 weeks in the ice-cube compartment or for 1 year in the freezer.

Hard-boiled eggs will stay up to 10 days in the refrigerator, although the whites begin to get rubbery after about 2 days.

Cooked-egg dishes should not be kept at room temperature for more than 30 minutes, but may be stored up to 4 days in the refrigerator, 3 to 4 weeks in the ice-cube compartment, or 3 to 4 months in the freezer.

Unopened packages of dried eggs can be kept at room temperature for up to 4 months. Once the packet is opened, dried eggs should be stored in a closed container in the refrigerator, where they will keep up to 1 year. Use only in cooked dishes unless the label specifies otherwise.

Dairy Products

MILK AND CREAM

Milk, cream, and chocolate or other flavored milk drinks will last up to 5 hours at room temperature, 3 to 5 days in the refrigerator. Milk or liquid cream does not freeze well; however, whipped cream may be stored in the freezer for a few weeks.

Buttermilk can be kept up to 2 weeks in the refrigerator, though it is best to use as soon as possible since the acidity increases during storage.

Cans of condensed and evaporated milk can be kept for 1 year, unopened, at room temperature. Turn the cans upside down every few months to prevent formation of lumps in the milk. Once opened, the canned milk can be kept up to 4 days at room temperature or 10 days in the refrigerator.

Powdered milk and cocoa mix can be kept 1 year at room temperature.

Sour cream lasts 10 days in the refrigerator.

Yogurt will keep for 2 weeks.

BUTTER

Covered or wrapped, butter can be kept 1 day at room temperature, up to 30 days in the refrigerator, 2 to 3 months in the ice-cube compartment, and 6 to 8 months in the freezer. Salted butter has a longer life-span than sweet butter.

ICE CREAM

Ice cream can be kept for 2 to 3 days in the ice-cube compartment and for up to 8 months in the freezer. Ice-cream cakes and sandwiches will last up to 4 months in the freezer; sherbets and ices, 2 months. Powdered ice-cream mix can be kept 1 year at room temperature.

CHEESE

Cheese will not turn rancid, but will become stronger with age until it is no longer palatable. Mold on the outside of cheese is not physically harmful; but should be cut or scraped off. The rind or wax coating on cheese will protect exposed surfaces. Once the cheese has been stripped of such cover, you can keep it moist and preserve its consistency by buttering the exposed edges and wrapping the cheese in plastic. You can assume with reasonable certainty that the harder the cheese, the longer its life-span.

Tightly wrapped, extremely hard cheeses can be kept up to 4 months at room temperature, 3 to 8 months in the refrigerator, up to 2 years in the freezer. Wrapping hard cheeses in damp cloths will provide necessary moisture and thus prolong their lives.

Soft cheeses can be kept up to 3 days at room temperature, 1 to 2 weeks in the refrigerator, 2 months in the ice-cube compartment, and 1 year in the freezer. Ripening cheeses like Camembert, Brie, and Liederkranz will continue ripening in the refrigerator; they should be left in their original containers until a few hours before serving.

Freezing tends to cause cheese to crumble, but several types of cheeses freeze fairly well, among them cheddar, Swiss, Edam, Gouda, brick Muenster, Port du Salut, provolone, mozzarella, and Camembert. Avoid freezing cottage cheese, cream cheese, or other processed cheeses, for they become watery when thawed.

Unopened cheese spreads can be kept at room temperature up to 1 year. Once opened, they will last up to 10 days at room temperature; up to 3 weeks in the refrigerator.

Cooked cheese dishes, like rarebit or fondue, may be kept for 1 day at room temperature, 3 days in the refrigerator, 1 month in the ice-cube compartment, or 1 year in the freezer.

Vegetables

Vegetables deteriorate nearly as fast as fish and shellfish. Almost any vegetable loses much of its subtle flavor within hours after it is picked. Natural sugars begin turning to starch the instant a plant is cut from its root; and volatile esters, the most important flavoring chemicals in many vegetables, begin to evaporate immediately. Anyone who has ever tasted one of the sweet varieties of cabbage straight from the garden knows how great a difference the disappearance of these ingredients can make. One of the best arguments for the use of some frozen vegetables is that they are blanched and frozen before this loss is quite complete. To minimize vitamin loss, frozen vegetables should not be defrosted, but should go directly from freezer to pot.

Though a canned vegetable rarely tastes as fresh as the frozen product, for many vegetables canning is a more effective storage technique than freezing because the shelf life is sometimes considerably longer. Beans, for example, last about 10 months in the freezer, but up to 3 years when canned. When using canned vegetables, however, keep in mind that much of their nutrient value is found in the canning liquid, not in the vegetables themselves.

Remember also that while many vegetables can be kept in the refrigerator for a week or more without loss of flavor or texture, their nutrient value slowly diminishes. Even the most storable vegetable is better for you if it is eaten soon after it is picked.

TRUFFLES

Truffles are unquestionably the most exotic and expensive of all vegetables. They grow only in a few areas of the world at the roots of certain trees, and are harvested once a year by specially trained pigs or dogs. The warty black, brown, or beige fungi taste like oysters, garlic, or Brie—depending on your palate.

Beige truffles should be kept under refrigeration and covered with sherry; that way they will stay tasty for about 1 week before they start to go soft. Black truffles last perhaps 2 or 3 weeks longer than beige truffles. They too should be kept covered and refrigerated, but they do not require the sherry. Both black and beige truffles can be frozen: Slice them and freeze the slices in separate pieces of foil so they can be used one at a time. In the freezer, they will last up to 1 year.

If only part of a can of truffles is used, be sure to cover the rest tightly and refrigerate, since the flavor and aroma quickly deteriorate once the can is opened. If the truffles remaining in the open can are not going to be used for some time, pour bacon or poultry drippings over the truffles, cover tightly, and keep in the refrigerator.

STORAGE LIFE-SPANS FOR VEGETABLES

Vegetable	Raw, on refrigerator shelf (days)	Cooked or canned, opened, on refrigerator shelf (days)	Life-span Prepared for freezing, in ice-cube compartment (months)	Prepared for freezing, in freezer (months)	Canned and unopened, on kitchen shelf (years)	Special instructions for storage
Artichokes	Up to 4	4–5	1	12	1	Keep cool and moist in closed container or plastic bag.
Asparagus	Up to 4	4	1	12	1–2	Wrap cut ends in damp paper towels; revive limp asparagus by cutting slices off ends of stalks and standing asparagus in cold water.
Beans: green beans, pole beans, wax beans	3–5	1–4	1	10	3	Store unwashed in plastic bag.
Beans: dried*	1 year (kitchen shelf)	1–4	Cooked: 1	Cooked: 4–6	——	Once opened, store in clean, covered container on kitchen shelf.
Beans: lima	In pod: 3–14	1–4	Fresh: 1 Prefrozen: 3	Fresh: 10 Prefrozen: 12	1	Store unshelled in moisture-proof container.
Beets	Up to 30	1–4	1	9–12	1	Brush off dirt, but don't wash; store loosely in plastic bag.

						Special Handling
Broccoli	3–5	1–4	1	6	—	Store in plastic bag.
Brussels sprouts	Up to 4	1–4	1	11	—	Store wrapped in original container and wash only just before cooking.
Cabbage	3–8	1–4	Don't freeze	Don't freeze	Don't freeze	Rinse head, drain dry, then wrap in plastic.
Carrots	Up to 30	4–5	2–3	12	1	Store in plastic bag.
Cauliflower	3–5	1–4	1	11	—	Store in plastic bag.
Celery	3–8	4–5	Don't freeze	Don't freeze	Don't freeze	Store leaves in plastic bag and stalks in foil (don't separate stalks from root); can be freshened by cutting a slice off stalk and standing celery in cold water.
Corn	Up to 4	1–4	1	12	3	Can be refrigerated for a few days but loses flavor quickly, so eat as soon as possible; store unhusked in plastic bag.
Cucumbers	7–14	1–2	Don't freeze	Don't freeze	Don't freeze	Don't peel or slice until ready to use; once cut, cover exposed ends with plastic wrap.

*Includes black or turtle beans, blackeye and yelloweye peas, chick peas (garbanzo beans), lima beans, Mexican chile beans (red beans), pinto beans, red kidney beans, soybeans, and all types of white beans.

STORAGE LIFE-SPANS FOR VEGETABLES

Vegetable	Raw, on refrigerator shelf (days)	Cooked or canned, opened, on refrigerator shelf (days)	Life-span Prepared for freezing, in ice-cube compartment (months)	Prepared for freezing, in freezer (months)	Canned and unopened, on kitchen shelf (years)	Special instructions for storage
Dandelion greens	3–8	4–5	2–3	12	—	Discard damaged leaves and roots; wrap in plastic and refrigerate in vegetable compartment.
Eggplant	3–14	4–5	2–3	12	—	Refrigeration not necessary, but keeps best in a cool, humid place; should be used within a few days, for it shrivels and goes limp rapidly.
Endive	3–8	—	Don't freeze	Don't freeze	—	Store in moisture-proof bag; remove and discard bruised leaves; thoroughly wash in cold water, shake off excess water, and blot dry with paper towel.

Garlic	Don't refrigerate	—	Don't freeze	Don't freeze	Kept in a tightly closed jar in a cool, dry place away from other foods, it will last several months; refrigeration will hasten spoilage.
Greens: for salads† / for cooking‡	3–8 / 3–8	1–4 / —	Don't freeze raw greens 2–3 (cooked)	Don't freeze raw greens 1 (cooked)	Trim, remove discolored leaves, wash thoroughly, drain, blot dry with paper towels, store in plastic bag; don't freeze raw greens.
Kohlrabis	Up to 14	1–2	2–3	12	Member of cabbage family, with leaves similar to those of turnips; best to store in a cool, dry place with good air circulation rather than in refrigerator.
Leeks	3–8	1–2	2–3	12	Don't wash until ready to use, but remove rootlets and unusable part of tops, leaving about 2 inches of green leaves; store in plastic bag in refrigerator.

†Lettuce, chicory, endive, escarole, watercress.
‡Beet tops, collards, dandelions, mustard greens, spinach, kale, Swiss chard, turnip tops.

STORAGE LIFE-SPANS FOR VEGETABLES

Vegetable	Life-span					Special instructions for storage
	Raw, on refrigerator shelf (days)	Cooked or canned, opened, on refrigerator shelf (days)	Prepared for freezing, in ice-cube compartment (months)	Prepared for freezing, in freezer (months)	Canned and unopened, on kitchen shelf (years)	
Lettuce	3–8	——	Don't freeze	Don't freeze	——	Rinse, dry, and store in plastic bag; head lettuce can be freshened by cutting a slice from the bottom and setting the head in cold water; the crisper the lettuce, the longer it will last. Never store lettuce near pears, plums, apples, tomatoes, or avocados, for the fruits give off a gas that can cause the lettuce to develop brown spots.
Mushrooms	Up to 7	4–5	2–3	12	1	Place on rack or shallow tray and cover with damp paper towels; make sure air can circulate around them.
Okra	4–5	4–5	1	12	1	Store in plastic bag in refrigerator.

Vegetable						Storage
Onions: green	Up to 7	4–5	Cooked: 2–3	12	1	Store green onions in plastic bag, others in mesh bag or container that permits air circulation; do not store near potatoes, which give off moisture and cause onions to sprout. Dry onions (yellow and red) can be stored outside refrigerator in a cool, dry place.
yellow, red, or white	7–28	4–5	Cooked: 2–3	12	1	
Parsley	3–8	—	2–3	12	—	Wash, shake off excess water, and store in plastic bag or tightly closed container.
Parsnips	Up to 30	2	2–3 weeks	12	1	Store in plastic bag in refrigerator.
Peas	In pods: 3–5	4–5	2–3	12	1	Store unwashed in plastic bag; do not shell until ready to cook.
Peppers	4–5	1–2	2–3	12	1	Wash, dry, and store in plastic bag in refrigerator.
Potatoes (white)	Except for new potatoes, do not refrigerate raw	4–5	2–3	12	1	Store in a cool (50° to 55° F), dark, dry, well-ventilated place; don't freeze uncooked potatoes.
Pumpkins	Whole: 30–120	4–5	2–3	12	1	Store in a cool, dark, dry place or refrigerate.

STORAGE LIFE-SPANS FOR VEGETABLES

| Vegetable | Raw, on refrigerator shelf (days) | Cooked or canned, opened, on refrigerator shelf (days) | Life-span | | Canned and unopened, on kitchen shelf (years) | Special instructions for storage |
			Prepared for freezing, in ice-cube compartment (months)	Prepared for freezing, in freezer (months)		
Radishes	Up to 7	—	Don't freeze	Don't freeze	—	Remove tops and rootlets, wash thoroughly, and store in plastic bag in refrigerator.
Rutabagas (Swedish turnips)	Up to 30	—	Don't freeze	Don't freeze	—	Store in a dry, well-ventilated place at about 55° F or refrigerate.
Scallions	Up to 6	—	Don't freeze	Don't freeze	—	Tops may get slimy if stored too long.
Shallots: fresh dry	1–2 60–90	—	Don't freeze	Don't freeze	—	Store in closed jar or plastic bag in refrigerator.
Spinach	3–5	4–5	3	12	1	Wash thoroughly, drain, and store in plastic bag in refrigerator.

Squash, acorn	Up to 30	4–5	2–3	12	1	Store in plastic bag in refrigerator. Purchase in small quantities.
Squash, Hubbard	120–180	4–5	2–3	12	1	Avoid having squashes touch each other.
Squash, summer	3–10	4–5	2–3	12	1	Store in plastic bag in refrigerator. Purchase in small quantities.
Squash, winter (butternut): unshelled	180	4–5	12	12	1	Store in a cool, dry, dark, well-ventilated (50° to 55° F) place.
shelled	30–90	4–5	12	12		
Sweet potatoes	Up to 30	4–5	2–3	12	1	Refrigerate or store at room temperature in a cool, dry place.
Turnips	30	2–3	2–3	12	1	Store in a cool (55°F), well-ventilated place.
Watercress	Up to 7	—	Don't freeze	Don't freeze	—	Wash, drain on paper towels, and store in refrigerator in tightly covered jar.
Yams	7	4–5	3	12	1	Store uncut in a cool, dark, dry place.
Zucchini (squash)	3–4	4–5	2–3	6–12	1	Store in plastic bag in refrigerator.

Fruits

For the most part, fruits seem to survive freezing and canning in better condition than vegetables do, perhaps because their high acidity protects them from damage. The nutritional value of canned fruit is relatively high, but nonetheless many of the vitamins leach out into the canning fluid. Because most canned fruit is packed in syrup, it may be better not to use the liquid and to accept the vitamin loss rather than to add unneeded sugar to your diet.

Slightly green fruits are often better buys than those that appear ripe in the store, and they may taste better in the long run. Bananas and some other fruit are "ripened" artificially and never quite develop the flavor of those that ripen on the tree or at home. Others, such as pineapples, will never ripen if picked too early and should be avoided unless clearly ready to eat. Fruits not meant to be eaten immediately should be selected with extra care, for even slight bruises or skin punctures can markedly reduce their storage lives.

Fruit should be ripened at room temperature, out of direct sunlight, then refrigerated. It will last longer if you don't wash the fruit until serving time.

One of the earliest methods of food preservation was drying fruit, a process that removes about 50 percent of the fruit's water content but almost none of its nutrients. Dried fruit should be stored in an airtight container at room temperature or in the refrigerator and will keep at least 6 months, often 1 year or even longer. If it becomes too dry, an hour's soak in water will restore its original quality.

STORAGE LIFE-SPANS FOR FRUIT

Fruit	Raw, at room temperature (days)	Raw, on refrigerator shelf (days)	Cooked or canned, opened and covered, on refrigerator shelf (days)	Prepared for freezing, in ice-cube compartment (months)	Prepared for freezing, in freezer (months)	Canned and unopened, on kitchen shelf (years)	Special instructions for storage
				Life-span			
Apples	60 to 240 (2–8 months)	Fully ripe: up to 14	—	2	12	—	Hard, freshly picked apples can be stored in a cold, humid place up to 8 months.
Apricots	Up to 14	Ripe: 3–5	2–4	2	12	2–5	Allow green apricots to ripen at room temperature
Avocados	1–3	14–28 (whole) 1–2 (cut and covered)	—	2	12	—	Place unripe fruit in paper bag on top of refrigerator to hasten ripening; do not refrigerate until ripe.

STORAGE LIFE-SPANS FOR FRUIT

Fruit	Raw, at room temperature (days)	Raw, on refrigerator shelf (days)	Cooked or canned, opened and covered, on refrigerator shelf (days)	Prepared for freezing, in ice-cube compartment (months)	Prepared for freezing, in freezer (months)	Canned and unopened, on kitchen shelf (years)	Special instructions for storage
				Life-span			
Bananas	Up to 10	Ripe: 3–5	—	1	12	—	If green or slightly green, allow to ripen up to 5 days at room temperature. Ripe bananas *can* be refrigerated—the skin will darken but the fruit is still good—store uncovered.
Berries: blueberries, blackberries, boysenberries	1–2	Up to 2	2–4	2	9–12	1	Do not wash before storing.
cranberries	1–4	7–28	14–30	2–3	12	1	
gooseberries	1–2	Up to 2	2–4	2	12–18	1	
huckleberries	1–2	7–14	—	2–3	12	1	
raspberries	1–2	1–2	—	2–3	12	1	
strawberries, loganberries	2–3	2–3	4–5	2–3	12	1	
Cherries	3–7	3–14	Fresh: 4–5 Maraschino (tightly covered): 60	2	12	Light cherries: 3 Dark cherries: 1	Keep unwashed in plastic bag in refrigerator.

Coconut	Whole: 60–180 (2–6 months)	Cut and wrapped: 7 Coconut milk: 7	30	2–3	Fresh: 9 Dried: 6	1–5	Store uncut coconut at room temperature. Store pieces of cut coconut or grated coconut tightly covered; if too hard, heat over hot water.
Figs	Dried: 180–240	Up to 3	4–5	Don't freeze	Don't freeze	1	
Grapes	2–3	14	4–5	Juice: 3 weeks	Juice: 12	1	Keep unwashed in plastic bag; avoid grapes with dry, brittle stems and those whose general appearance is shriveled or sticky.
Grapefruits	Uncut: 2–7	Uncut: 30–120 Fresh juice: 1–2	5	Juice: 3 weeks	12	1	Avoid extremes of temperature.
Kiwi berries (Chinese gooseberries)	Up to 7	—	—	2	12	—	Allow to ripen at room temperature for 3–5 days. Place in plastic bag and store at room temperature.
Lemons	Fresh juice: 1	Uncut: up to 28 Fresh juice: 5–6	Juice: 1–4	Juice: 3 weeks	Juice: 12	1	

STORAGE LIFE-SPANS FOR FRUIT

Life-span

Fruit	Raw, at room temperature (days)	Raw, on refrigerator shelf (days)	Cooked or canned, opened and covered, on refrigerator shelf (days)	Prepared for freezing, in ice-cube compartment (months)	Prepared for freezing, in freezer (months)	Canned and unopened, on kitchen shelf (years)	Special instructions for storage
Lemonade	—	7	—	3 weeks	12	—	
Limes	Fresh juice: 1	Uncut: up to 28 Fresh juice: 7	Juice: 1–4	Juice: 3 weeks	Juice: 12	1	
Mangoes	—	7	7	2–3	12	—	Wrap in waxed paper and refrigerate.
Melons (cantaloupes, honeydews)	3–5	Whole: 4–8 Cut and covered: 3–4	—	1	9	—	Allow to ripen at room temperature. Cover cut surfaces with plastic wrap.
Nectarines	Up to 7	14	4–5	2–3	12	1	Allow to ripen at room temperature. The fruit is ripe when slightly soft along the seam and of good color (orange-yellow to red).
Oranges	Up to 14	30–60	5–6	Juice: 3 weeks	Juice: 12	1	Store in a cool, dry, well-

Fruit						Storage notes
Peaches	Up to 7	3–14	4–5	2–3	12	1 — Store unwashed and uncovered.
Pears	Up to 7	3–14	4–5	Don't freeze	Don't freeze	1 — Keep underripe fruit in cool, fairly humid place.
Persimmons	—	1–2	—	2–3	12	— For best ripening, keep in a cool, dark, dry place.
Pineapples	2–3	Up to 20	4–5	3	12	1 — Allow to ripen at room temperature, but not in sunlight.
Plums	Up to 7	14–21	4–5	2–3	12	1 — Allow to ripen at room temperature
Rhubarb	—	Up to 4	4–5	2–3	12	1 — Keep in plastic bag in refrigerator.
Tangerines	Up to 7	30–60	5–6	Juice: 3 weeks	Juice: 1	1 — Allow to ripen at room temperature. Wrap cut surfaces in waxed paper and refrigerate.
Tomatoes	Up to 14	Slightly underripe: 8–12 while ripening; Ripe: 2–3	4–5	In cooked dishes: 2–3; Don't freeze raw	In cooked dishes: 12	1 — Store unwashed and uncovered; allow green tomatoes to ripen at room temperature.
Watermelons	3–4	Up to 14	—	—	—	—

NUTS

Most nuts will remain fresh in their shells for 1 year; pecans and Brazil nuts, however, should be kept only for 6 months unless they are refrigerated. Exposure to light, air, warmth, and moisture causes nutmeats to become rancid, so once shelled, nuts should be used within 2 to 3 months, if stored in a tightly closed container and kept at room temperature, or 4 to 6 months if also refrigerated. Vacuum-packed cans of nuts remain fresh for 1 year, unopened. Once opened, they should be kept in the refrigerator.

Baked Goods

Homemade baked goods, as well as those from the local bakery, should be consumed no later than the day after baking. The products carried in supermarkets, on the other hand, will last anywhere from 3 days to 1 week, depending upon the kinds and amounts of mold inhibitors and preservatives they contain. Generally, the less moisture a baked product has, the less stable it is and the shorter its life-span. Bread made without oil or fat (French or Italian bread) begins to dry out after just a few hours. On the other hand, honey and molasses allow bread to remain fresh longer by improving its moisture retention.

Frozen bakery products have indefinite life-spans if they are kept tightly wrapped and sealed and are stored at 0° F. Frozen goods stored at temperatures higher than 0° F (in a self-defrosting freezing compartment, for example) have substantially shorter storage lives.

BREADS AND ROLLS

Baked bread, rolls, muffins, and biscuits can be kept up to 5 days at room temperature, 2 weeks in the refrigerator, 1 month in the ice-cube compartment, 3 to 6 months in the freezer. (Thawed bread becomes stale much faster than fresh.) Storing bread in the refrigerator may inhibit the growth of mold, but the cold makes bread dry out very quickly; refrigerate bread only when hot weather demands it. All bread should be kept in a closed container or wrapped in moisture-proof paper.

Brown an' Serve rolls have life-spans of up to 1 week in the refrigerator and can be frozen indefinitely.

CAKES AND COOKIES

Homemade cakes and cookies without fruit, nuts, fillings, or toppings can be kept for 1 to 2 weeks at room temperature or 6 months to 1 year in the freezer. Store soft cookies separately from crisp cookies in a tightly covered container in a cool place. A piece of apple or a slice of fresh bread in the cookie container will provide moisture if the cookies dry out. Store crisp cookies in a loosely sealed container. If they soften, heat the cookies in a slow oven for a few minutes to restore crispness.

Packaged cookies are protected with preservatives and will last up to 4 months at room temperature. Cakes and cookies with fruit and/or nuts can be kept up to 3 months at room temperature, 6 months in the refrigerator, or up to 1 year in the freezer. Fruitcakes fare best when wrapped first in cheesecloth soaked in rum or fruit juice, then in foil. Cakes and cookies with fillings or toppings (especially custards) will last only 3 to 4 hours at room temperature or 2 days in the refrigerator.

PIES AND PUDDINGS

Fruit pies can be kept for 2 to 4 days at room temperature, 3 to 7 days in the refrigerator, 1 month in the ice-cube compartment, or (if unbaked) 6 to 8 months in the freezer. Baked custard pies, such as pumpkin pies, last about the same time as fruit pies, but soft custard pies (or pies filled with cream or pudding) should be kept only 1 day at room temperature, 2 to 3 days in the refrigerator, and 2 months in the freezer. Unbaked piecrusts will last 2 days at room temperature, 7 to 10 days in the refrigerator, 1 month in the ice-cube compartment, or 6 to 8 months in the freezer. Baked piecrusts will last 1 to 4 days at room temperature or 3 to 7 days in the refrigerator. Puddings such as bread pudding and rice pudding can be kept 4 to 7 days in the refrigerator, 1 month in the ice-cube compartment, or 6 to 8 months in the freezer.

DOUGH

Unbaked bread dough can be kept up to 2 days at room temperature, 5 days in the refrigerator, 2 weeks in the ice-cube compartment, or 6 months in the freezer.

Yeastless frozen bread dough can be kept in the freezer from 5 to 10 weeks (the ice-cube compartment isn't cold enough). Yeast dough will keep in the refrigerator for 2 weeks (but it is preferable to freeze yeast bread or rolls after baking). Once removed from the freezer, the dough should be used within 24 hours. After that time, the yeast, which remained inactive during freezing, will become partly active while the growth of substances upon which the yeast feeds will remain inhibited. As a result, the yeast will literally starve to death.

Unbaked cookie dough can be kept 3 days at room temperature, 3 weeks in the refrigerator, 3 months in the ice-cube compartment, and 6 months in the freezer.

Pasta, Grains, Mixes, and Baking Ingredients

PASTA AND RICE

Pasta (macaroni, noodles, spaghetti, and so forth) and rice last up to 2 years in a cool, dry place if kept unopened in the original packages or in tightly closed containers, 3 to 6 months if opened. Once cooked, they can be kept 4 to 5 days in the refrigerator, 3 to 4 weeks in the ice-cube compartment, or 1 year in the freezer.

CEREALS

Cereals that require cooking—cream of wheat and oatmeal, for example—can be kept up to 6 months in tightly closed containers. Whole-grain cereals should be refrigerated even when they haven't been opened because of their fat content; they will keep 5 to 6 months uncooked. Cooked cereals will keep 2 to 4 days if refrigerated. Unopened ready-to-eat cereals will keep on the kitchen shelf for about 4 months. Their shelf lives are shortened considerably if they are stored in a damp place.

FLOURS

Stored at room temperature, white flour can be kept 1 year in its original package or in a tightly closed container. Weevils and flour bugs that occasionally infest flours are not harmful and can be sifted out. Wheat germ and dark flours (such as rye and soybean flours) spoil more quickly and should be stored in the refrigerator, where they will keep up to 6 months, or in the freezer, where they last up to 1 year. Potato flour deteriorates quickly and shouldn't be stored for more than 1 to 2 months.

LEAVENING AGENTS

Baking soda and baking powder, as well as cream of tartar, can be stored for 6 to 8 months. They must be kept tightly covered to prevent moisture absorption.

Active dry yeast should be stored on a cool kitchen shelf until the expiration date indicated on the package. Compressed yeast should be kept in the refrigerator until the expiration date—1 or 2 weeks. It may be stored in the freezer for several months. If it crumbles easily, chances are it is still good.

PANCAKE, WAFFLE, AND CAKE MIXES

These convenience foods will last up to 3 months at room temperature or 5 to 6 months in the refrigerator. Uncooked batter will last no longer than 1 day at room temperature, but can be left up to 4 days in the refrigerator.

Oils and Fats

The shelf lives of various oils and fats used in cooking vary considerably, as the accompanying table indicates. (Gourmands should be aware that excessive use of animal fats in cooking will be detrimental to their own life-spans.)

Oils should be tightly covered and stored in a cool, dry place, since they fade when exposed to light and turn rancid if exposed to moisture. Refined oil turns rancid faster than unrefined, so olive oil outlasts other oils pressed from vegetables and nuts—and keeps better in cans than in bottles. Refrigeration may cause oils to cloud, but they will become clear again when returned to room temperatures. Olive oil, however, may solidify if refrigerated too long.

STORAGE LIFE-SPANS FOR OILS AND FATS

Type of oil or fat	At room temperature	On refrigerator shelf (months)	In ice-cube compartment (months)	In freezer (months)
		Life-span		
Margarine	1 day	3	3	6–8
Solid vegetable shortening	2–4 months	Don't refrigerate	Don't freeze	Don't freeze
Lard	Up to 2 days	2	4–5	12
Bacon or chicken fat	Up to 24 hours	Up to 1	2–3	6–8
Cooking and salad oils	Up to 4 months	Up to 12	Don't freeze	Don't freeze

Herbs, Spices, and Condiments

HERBS, SPICES, AND EXTRACTS

The aroma and flavor of herbs derive primarily from the oil contained in their leaves and seeds. Dried herbs, if kept away from heat and light, will last up to 6 months in tightly closed containers. Fresh herbs should be blanched before freezing or placed in an ice-cube tray partially filled with water (and transferred to plastic bags once frozen). They can be frozen for 6 months; but better yet, grow your own and harvest as needed.

Spices are made from the pungent roots, barks, stems, buds, fruits, leaves, or seeds of various tropical or subtropical plants.

Whole spices will retain their flavor for up to 1 year; ground spices lose their potency after 6 months. Some spices, like chili powder and ground red pepper, are prone to infestation by small insects and should be checked every few months.

Extracts can be stored for at least 6 months.

SALT

Salt has a life-span of up to 2 years at room temperature (better yet, near the stove) in a closed container. If the salt begins to cake, it can be dried out in a warm oven.

PEPPER

Whole peppercorns can last for decades, but once they are ground they quickly lose their potency and should be stored no more than 1 month.

KETCHUP AND OTHER ACCOMPANIMENTS

Ketchup, relishes, pickles, and chutneys can be stashed away in a cool dark place for a year or more. Once the bottle or jar is opened, store in the refrigerator for 2 to 3 months.

Flavor intensifiers such as Tabasco and Worcestershire

sauces will keep up to 2 years tightly capped on the kitchen shelf. Keep in mind when buying these and other condiments that along with the convenience and long shelf lives of store-bought condiments you are also getting large doses of preservatives—sodium benzoate, polysorbate, calcium chloride, and gum tragacanth, to name a few—as well as stabilizers and firming agents. You may consider this an unacceptable trade-off.

HORSERADISH

The life-span of processed horseradish depends less on its age than how well it is refrigerated. In the refrigerator, horseradish should last from 6 to 9 months; outside, during a hot spell, it will begin to lose its strength almost immediately and will be spoiled even within 48 hours.

When horseradish goes bad, you'll know it. Despite the presence of vinegar—a natural preservative—horseradish is relatively volatile and unstable. Its demise is a result of oxidation, a chemical process that changes the normal cream color first to brown and then to a deep black.

MAYONNAISE

Mayonnaise is a very stable product that, according to the research laboratory for Best Foods (Hellman's mayonnaise), will last indefinitely if the jar remains unopened. Nevertheless, mayonnaises generally do not stay on the grocer's shelf for more than 9 months; after that time, the flavor begins to go and the mayonnaise loses its freshness. After the jar is opened, mayonnaise should be tightly covered and refrigerated; then it should last about 2 months.

MUSTARD

Prepared mustard contains loads of vinegar (and often propylene glycol alginate), which protects it from spoiling. Even after it is opened, mustard stays pretty much as good as new. A jar can remain indefinitely on the shelf, but it's best to refrigerate homemade mustard and opened jars of commercial mustard. Although mustard may separate, simple stirring blends it again. The only minor limitation to the shelf life of mustard is that, after a year or so, its spices may lose some of their bite; but the difference in taste is minimal.

SALAD DRESSING

Bottled salad dressing can be stored for 6 months in the refrigerator (once opened) or at room temperature. If the flavor begins to deteriorate, the oil has gone rancid, and the dressing should be thrown out.

VINEGAR

Cider or distilled vinegar will keep indefinitely, unopened, in a cool, dark place; up to 6 months, once opened. Other types can be stored for 6 months. Sediment is not harmful, but if mold appears the vinegar has spoiled. Sometimes vinegar develops a jellylike membrane called "mother"; it is not harmful and can be removed by straining the vinegar through a fine sieve.

Other Perishables

SOUPS, STOCKS, AND GRAVIES

These liquids will last no more than 1 day at room temperature, up to 4 days in the refrigerator, 2 to 3 months in the ice-cube compartment, 1 year in the freezer, and 1 year on the kitchen shelf (canned). Gravies with high fat content are unsuitable for freezing unless the fat is skimmed off first. To extend the life-spans of soups and gravies, bring to a boil and simmer for 10 minutes every few days; they then can be kept almost indefinitely.

Bouillon cubes and soup mixes can be kept 1 year or more in a dry, cool place.

SAUCES

As a rule, egg or cream-thickened sauces keep poorly. They should be left out no more than 30 minutes at room temperature, but can be refrigerated for 4 days. They do not freeze well. Starch-thickened sauces will last up to 12 hours at room temperature, up to 3 days in the refrigerator, 2 to 3 months in the ice-cube compartment, and 1 year in the freezer.

Acidic sauces, especially tomato sauces, are less perishable, and may be safely refrigerated for 2 weeks. They can be stored in the ice-cube compartment for 2 to 3 months or for 1 year in the freezer. Canned sauces may be kept on the shelf for 1 year.

Dessert sauces—chocolate, hard sauce, butterscotch sauce, and so forth—can be kept at room temperature for up to 30 days or in the refrigerator up to 6 months. They do not freeze well.

Sweet Pleasures

SUGAR

White sugar, if tightly covered, can be kept at room temperature indefinitely. Brown sugar and confectioner's sugar will last up to 4 months on the shelf, but to retain maximum moisture they should be stored in plastic bags in the refrigerator.

MAPLE SYRUP, CORN SYRUP, AND FRUIT SYRUPS

Unopened, these syrups will last at least 1 year on the shelf. Once opened, they will keep up to 10 days at room temperature or 2 months in the refrigerator. Crystallization may occur at low temperatures, but a little heat will dissolve the crystals. Formations of mold are harmless and can be scraped off.

MOLASSES

Unopened, molasses will keep at least 2 years; once opened, it can be stored for 2 months in the refrigerator.

HONEY

Honey keeps a year unopened or several months at room temperature. Make sure it is tightly covered; otherwise it will absorb moisture and ferment. Refrigeration will cause honey to become grainy.

JELLIES AND JAMS

Unopened, jams and jellies last 1 year or more at room temperature. Once opened, they will keep 1 to 2 weeks at room temperature, 4 to 5 weeks in the refrigerator, 6 to 8 months in the ice-cube compartment, or 1 year in the freezer. Jellies and jams made without cooking *must be* refrigerated; they will keep 3 to 4 months in the refrigerator and 1 year in the freezer.

CHOCOLATES

Mass-market, mechanically produced chocolates like Nestlés and Hershey's can last from 9 to 15 months in bars and 15 months in morsels, if they are stored at a constant room temperature. There are two reasons for their long shelf lives: (1) they are often loaded with preservatives and (2) they are actually made so that they "ripen" in the 6 months or so they spend in the warehouse before being delivered to retail outlets.

Mechanically made commercial chocolates produced by smaller manufacturers who own their own outlets, like Barton or Barricini, have shelf lives of 3 to 4 months, if the temperature and humidity remain constant at 68° to 70° F and 50 to 60 percent, respectively. The filled bars of these chocolates contain preservatives; the solid chocolates do not.

Still smaller manufacturers that produce "homemade" chocolates estimate the shelf lives of their products at a maximum of 4 to 6 weeks. And some confections, like the hand-dipped strawberries of Krön Chocolatier in New York have suggested shelf lives of only 1 day.

If chocolate becomes too warm, the fat in it rises to the surface, causing a grayish film to appear. When this happens the chocolate is still edible; if the wrapper becomes oily, however, the chocolate has deteriorated.

Baking chocolate, dry cocoa, or hot-chocolate mix will last 1 year at room temperature if kept in a tightly closed container.

HARD CANDIES

Hard candies and caramels keep best if individually wrapped in foil or plastic and will last 3 to 12 months at room temperature. If they contain fruit or nut fillings, they will keep only 3 to 4 months unless refrigerated. There they'll stay reasonably fresh for up to 8 months.

MARZIPAN

When it is stored in an airtight container and refrigerated, marzipan will keep for several weeks.

JELL-O

Packaged Jell-O and plain gelatin keep indefinitely in a cool, dry place, if unopened. Prepared Jell-O keeps 5 to 7 days or longer, but it becomes watery after a while. Gelatin desserts cannot be frozen.

Beverages and Beverage Bases

COFFEE

Green coffee beans—if they are stored so that they cannot absorb moisture or strong odors—may be kept for several years.

In fact, the flavor of certain green coffee beans may actually be enhanced by a little aging.

Roasted beans may be kept for 3 to 4 weeks, if they are stored in a glass jar in the refrigerator. The low temperature retards their going stale and the glass protects them from foreign odors. Beans can also be stored in the freezer for a few months (protected from moisture and odors) and can be ground and brewed without first being defrosted.

Ground coffee will keep in the refrigerator (in a glass jar) for 7 to 10 days and in the freezer for a month.

Vacuum-packed ground coffee—in sealed cans—keeps fresh for up to 3 years, if stored unopened at room temperature (70° F).

Mass-produced instant coffee can be stored, unopened, at room temperature, for at least 2 years.

Freeze-dried coffee is made by flash freezing fresh coffee, then drawing off the ice crystals by a vacuum process that does not affect the flavor as much as heat dehydrating does. It can be stored, unopened, for at least 1 year. Once instant or freeze-dried coffee has been opened, it should be used within 2 weeks for maximum flavor retention. In addition, it should be refrigerated in a tightly closed jar.

TEA

Unopened, loose tea and tea bags will keep up to 18 months before becoming stale. Tea should be stored in a closed container in a cool, dry place, away from foods whose flavors it might absorb.

COCOA

Dry hot-cocoa mixes last about a year, whether their cans are opened or remain sealed. However, excessive heat, constant opening and closing of the can, or sticking a wet spoon in the powder may cause the mixture to cake. In addition, extensive exposure may cause some loss of flavor.

FRUIT JUICES

Fruit juices, be they fresh, canned, or reconstituted from frozen concentrates, have refrigerator life-spans of 5 to 7 days.

CARBONATED DRINKS

Sodas can be kept indefinitely if unopened. After opening, they will last about 1 day at room temperature, 10 to 14 days in the refrigerator, if tightly stoppered.

BEER

Even if properly refrigerated, beer cannot be stored indefinitely. The recommended maximum is 3 months, after which the flavor can begin to deteriorate, developing a buttery or papery taste. Beer should not be exposed to light—the reason most beers come in dark bottles or cans.

WINE

The life-spans of wines vary drastically, and determinations are extremely complex. Some wines (reds as well as whites) *should* be consumed when very young; they are best when less than 1 year old. Others take years, even decades to mature. A number of wines are over 100 years old and still in excellent condition. The record goes to a bottle of Steinwein 1540, salvaged from the cellar of King Ludwig of Bavaria. When opened, the wine was over 400 years old and still drinkable.

For ideal storage, bottles of wine should be kept on their sides in a cool (50° to 60° F), dark place undisturbed by either fluctuations in temperature or unnecessary handling. Once opened, a bottle of fine wine should be consumed in its entirety. White table wines can be recorked and kept for a day or so, reds somewhat longer; then they quickly turn to vinegar. Sweet wines can be kept 1 week after being opened and then can still be used for cooking. Fortified wines will last a few weeks after opening—with some deterioration. Aromatic wines, however, keep well for several weeks.

Whiskey

If kept unopened and away from heat and light, distilled spirits have indefinite life spans. However, some spirits do have optimal ages, when they are at their best. Scotch, for example, originally remained unblended and was called malt whiskey. But malt whiskey was found to deteriorate rapidly. Today, what most of the world refers to as Scotch is a blend of up to 60 percent malt whiskey and other spirits distilled from corn, rye, or oats. It matures in wooden casks and becomes more expensive as it gets older. Most experts feel, however, that Scotch does not improve after its twelfth year. Once bottled, Scotch will not lose its quality as long as it is firmly stoppered.

In contrast, the optimal age for Bourbon is between 8 to 12 years; for rye, 4 to 6 years. Rum is more controversial: Some people believe rum does not improve with age; others claim it does and is best when it is 15. Gin and vodka do not improve with age.

Tobacco Products

CIGARETTES

Once opened, a package of cigarettes will last only a few days before the tobacco dries out appreciably. Unopened packs can be stored for longer periods, but they too will dry out over a period of months if stored in a dry place. If kept at about 60° F and 68 to 72 percent humidity, however, cigarettes, either opened or unopened, can last for decades.

Once a cigarette is lit, it has a more definite life-span. Commercially packaged cigarettes in the United States are rolled in a specially treated paper that continues to burn, even if the cigarette is not puffed. A regular-size, nonfilter cigarette, for example, will burn down completely in a little over 16 minutes if it is lit and simply left in an ashtray.

CIGARS

In the humidors of Alfred Dunhill, the respected London tobacconist, are boxes of cigars that are more than 40 years old, maturing to a fine color and flavor. The humidors at Dunhill's are kept at 60° F, 68 to 72 percent humidity.

PIPE TOBACCO

Vacuum-packed tins of pipe tobacco can be kept up to 20 years. Once opened, they can be stored for another 20 years under the right conditions—60° F, 68 to 72 percent humidity.

A well-tamped, well-tended pipeful of tobacco will burn for 25 to 40 minutes or even longer, depending on the smoker and the quantity and type of tobacco involved. In one pipe-smoking contest, Finn Yrje Pentikainen tucked a little over 3 grams of tobacco in his pipe and kept it going continuously for 4 hours, 11 minutes, and 28 seconds—a world record that the International Association of Pipe Smokers' Clubs has not yet seen fit to recognize.

5 / THE PRODUCTS OF MAN

"PEOPLE are fond of their cars," said David Riesman. "They like to talk about them—something that comes out very clearly in interviews—but their affection for any one in particular rarely reaches enough intensity to become long-term."

These days it seems that few of us generate sufficient affection for our belongings to keep them around for any length of time. We live in an age of impermanence.

There are several justifications for this throw-away mentality. As technology marches on, we are supposedly producing better and better products. Why build a widget to last 20 years when a better widget is sure to be available in a matter of months? For some items this notion holds true. The first pocket calculator, for example, introduced less than 10 years ago, cost upwards of $400 and was capable of doing only simple addition, subtraction, multiplication, and division. Today, calculators superior to these first models are available for under $5. For the cost of an early calculator you can now purchase a full-fledged computer. In the case of cars, on the other hand, Detroit turns out new models every year, and even spokesmen for the automotive industry would be hard put to explain why one year's model is better than the previous year's.

We are also told that, because of the introduction of automation in manufacturing, it is now cheaper to replace than to repair. As the energy crisis becomes more and more severe, this argument seems increasingly ridiculous.

Luckily, there are still master craftsmen among us—people who make things to last. High-quality, well-made products are still available. It is up to us to maintain them.

Structures and Their Materials

MONUMENTS

Man's urge to achieve immortality in his works is more visible in monumental architecture than in any of his other products, so it comes as no surprise that these edifices have exceptionally long life-spans. A large majority of the monuments that have survived into this century were designed for exactly that purpose and erected through efforts few other goals could command; Chartres cathedral, for example, took 66 years to build and the equivalent of over $100 million at a time when the city's population totaled a mere 10,000. With today's technology, it is theoretically possible to build structures that would endure virtually as long as the planet itself. That the ancients came as close as they did with only natural materials at hand is an impressive tribute to their need to create enduring as well as inspiring symbols expressive of their cultures.

Ironically, the same technology that today makes truly long-lived monuments possible may well destroy buildings that centuries of wind and weather left nearly untouched. Acid rains and the air pollution that causes them have begun to eat away historical landmarks all over the world. The problem is particularly acute in Paris, where high levels of atmospheric sulfuric acid have been eroding the stone of the Louvre and Notre Dame for several decades. To date, no effective way has been developed to shield these monuments from corrosion by the air.

AGES OF IMPORTANT MONUMENTS

Monument	*Year built*
Step Pyramid of King Zoser, Saqqara, Egypt	2686 B.C.
Great Sphinx, Giza, Egypt	2650 B.C.
Pyramid of Khufu, Giza, Egypt	2600 B.C.
Pyramid of Khafre, Giza, Egypt	2560 B.C.
Pyramid of Menkure, Giza, Egypt	c. 2525 B.C.
Palace of Minos, Knossos, Crete	c. 1500 B.C.
Tomb of Tutankhamen, Luxor, Egypt	c. 1325 B.C.
Temple of Poseidon, Paestum, Italy	c. 460 B.C.
Parthenon, Athens, Greece	448– 438 B.C.
Temple of Athena Nike, Athens, Greece	427– 424 B.C.
Monument of Lysicrates, Athens, Greece	334 B.C.
Stupa Number 1, Sanchi, India	3rd–1st centuries B.C.
Pont du Gard, Nîmes, France	Early 1st century A.D.
Colosseum, Rome, Italy	A.D. 72–80
Trajan's Column, Rome, Italy	106–113
Pantheon, Rome, Italy	c. 115–125
Basilica of Constantine, Rome, Italy	c. 310–320
Arch of Constantine, Rome, Italy	Early 4th century
Colossal Buddha, Yünkang, China	c. 450–500
Sant'Apollinare in Classe, Ravenna, Italy	c. 530–549
Hagia Sophia, Istanbul, Turkey	532–537
Hōryūji Monastery, Japan	7th century
Mosque, Cordova, Spain	785–990
Great Temple Compound, Bhuvaneshwar, India	8th–13th centuries
Chapel of Charlemagne, Aachen, West Germany	805
Temple of Quetzalcoatl, Mexico	9th century

Monument	Year built
Pagoda, Daigoji, Japan	951
Saint Mark's Cathedral, Venice, Italy	Begun 1063
Durham Cathedral, England	1093–1130
Temple of the Warriors, Chichén Itzá, Mexico	11th Century
Angkor Wat, Cambodia	Early 12th century
Notre Dame, Paris, France	1163–1250
Tournai Cathedral, Belgium	1171
Chartres Cathedral, France	c. 1194–1260
Salisbury Cathedral, England	1220–1270
Reims Cathedral, France	13th–15th centuries
Rouen Cathedral, France	13th–16th centuries
Inca Citadel, Machu Picchu, Peru	15th century
King's College Chapel, Cambridge, England	1446–1515
Saint Basil's Cathedral, Moscow, U.S.S.R.	1554–1560
Saint Peter's Basilica, Rome, Italy	1606–1660
Mosque of Sultan Ahmed I, Istanbul, Turkey	1609–1616
Taj Mahal, Agra, India	1630–1648
Louvre, Paris, France	Completed 1670
Palace of Versailles, France	1669–1685
Brandenburg Gate, Berlin, East Germany	1788–1791
Monticello, Charlottesville, Virginia, U.S.A.	1796–1806
Arc de Triomphe, Paris, France	Begun 1806
Houses of Parliament, London, England	Begun 1835
Trinity Church, New York, New York, U.S.A.	1846
Washington Monument, Washington, D.C., U.S.A.	1848
Lincoln Memorial, Washington, D.C., U.S.A.	1922

HOUSES

Under French law, all homes erected within the Paris city limits must be built to last through at least three generations of tenants. No such code exists in the United States, but architects here aim for a 50- to 60-year service life in their home designs. After six decades, the theory runs, there will be enough wrong with heating systems, plumbing, electrical wiring, and similar essentials to render the house uninhabitable even if the framing and walls remain intact. After all, the architects point out, even the colonial houses scattered across New England have had their roofs and siding replaced at least once a century. Add to that the extra cost of making a house more durable, the chance that technological advances and social change will markedly alter current life styles and the probability that population growth and inflation will make single-family dwellings a luxury only the wealthiest can afford, and a 60-year design life begins to seem long enough.

Building Materials

CONCRETE

The Colosseum of Rome was built 1,900 years ago. The Roman aqueducts have stood since the first century A.D. and are still in use. They are all made of concrete. Of all man's building materials, only stone has proved more durable, and stone may well hold its record only because it was used in construction millennia before deposits of natural cement were discovered.

Modern varieties of concrete are expected to prove even more enduring than those of Greece and Rome. Engineers planning the entombment of nuclear reactors have claimed that the dense, supremely resistant concrete chosen for such structures will still be standing long after the pyramids have crumbled to dust.

Today's concrete, a mixture of sand or gravel and portland cement hardened by a chemical reaction with water, was developed in the early nineteenth century. As a structural material, it is immensely strong compared with other nonmetals. Ordinary concrete will support compressive forces up to 3,000 pounds per square inch, and some specially formulated varieties will withstand nearly three times that force without crushing. Concrete's tensile strength, in contrast, is relatively low. To compensate for this, steel bars are often cast into concrete structures subject to tension. Alternatively, the beam can be compressed when it is cast so that the workload placed on it merely overcomes the built-in compression.

Unlike other materials, concrete grows stronger with age. (The water reaction that binds it together slows markedly as the concrete hardens, but it continues for decades.) And there are few natural forces that can damage a well-designed concrete structure. Yearly cycles of freezing and thawing that once caused the surface of concrete castings to flake off, broken down by ice crystals that formed inside the granular material, have been defeated by the creation of tiny air bubbles in the concrete that take up the expansion. And though reinforcing steel can rust, expanding and breaking down the surrounding concrete, this occurs only in poorly designed constructions.

WOOD

Fungi, termites, fire, and countless other destructive forces attack wood, particularly in warm, moist air. In hot, dry climes, however, wooden objects can last for surprisingly long times. One of the best examples is a 43-inch statute of Ka-aper found at Saqqara, Egypt. Sculpted around 2400 B.C., it remains in nearly perfect condition.

Even in less favorable climates, wood can remain sturdy for centuries. In frame houses, serious weakening is usually attributable less to simple aging than to attack by termites or carpenter ants or to faulty design that allows water to collect and cause rotting. Durability can be greatly increased by preservatives. Untreated railroad ties, for example, have life-spans of only 5 years or so. Creosoted ties last six times as long (see page 176). Processes developed in the last 15 years, ranging from irradiation to impregnation with acrylic plastics, make wood virtually indestructible, but most are so costly that they are seldom used.

BRICK

Brick can be nearly as indestructible as concrete. In the United States, however, one of the best proofs of the toughness of brick is barely a century old. Built in 1876, a Brooklyn, New York, apartment complex started its life as the first tenement building erected in this country. After decades of neglect that has left some apartments nearly uninhabitable, the buildings have been taken over by a new owner, who plans to turn them into middle-income housing. Though the apartments are being completely gutted for renovation, their brick underwalls are still sturdy enough to stand as they are.

Exposed brick walls do need some care. The lime mortar used to cement the bricks together in old walls needs replacing, or repointing, every 10 to 15 years in walls that face prevailing winds, every 30 to 40 years in sheltered walls. Modern cements last somewhat longer unless water penetrates the joints.

STEEL

Though tiny quantities of steel have been made for many hundreds of years, the metal has been used in construction only since the early 1800s. Widespread use of steel did not begin until the 1860s and 1870s, when the first commercial-scale steel-making processes were developed. Though rusting and corrosion can destroy unprotected steel in a few decades, architects claim that the steel frame of a skyscraper should remain safe indefinitely.

ALUMINUM

In pure form, aluminum is soft and not terribly strong. Alloyed with other metals, however, aluminum can be made with tensile strengths of up to 100,000 pounds per square inch—more than enough for use in almost any industry. Paradoxically, one of aluminum's great advantages over other building materials is the ease with which it combines with oxygen. Iron oxide is rust; it eats steel away, slowly destroying whatever it infects. Aluminum oxide, on the other hand, forms a thin coating on the metal's surface, protecting it from further oxidation and chemical attack. Anodizing aluminum to prevent weathering is simply an electrical process for making a thick oxide coating that penetrates into the metal itself. As a construction material, aluminum is expected to prove nearly immortal.

Housing Components

WALLS AND FRAMING

These are the slowest-deteriorating parts of any building. Barring fire or termites, no house should ever need major attention to its basic structure. At the end of a home's life-span, walls and frame should be virtually as good as new.

Inside wall surfaces are far less durable. The slight shiftings of a house as its foundation settles can cause plaster walls and ceilings to crack and flake in as little as 1 year. Sheetrock and paneling deteriorate more slowly, but neither is likely to remain attractive as long as half a century.

ROOFING

By far the longest-lived roofing materials are slate and copper terne, or sheeting. Properly made, a slate or copper roof can last well over a century—much longer, of course, than the intended life of today's home. A lot depends on the skill of the builder, however. A poorly made copper roof can be ripped apart when rainwater penetrates its seams and freezes. Both slate and copper are now so expensive that they are reserved almost entirely for churches and similar long-term structures.

The same arguments apply to tile roofing. Tile can last 75 years or more and is often found on homes built in the 1920s or 1930s. But rising costs have virtually eliminated it from modern houses.

Wood shakes and shingles are far less expensive, but not as durable. Shakes can last about 50 years, cedar shingles about 20. Again, however, it takes considerable skill to install them properly, and their popularity has dropped off markedly in recent years.

By far the most common roofing materials now in use are shingles made of felt, rag, or fiberglass and impregnated with asphalt. Though asphalt shingles survive only 15 to 25 years, their low cost and easy installation make it practical to replace them as needed. Asphalt shingles begin to break down when the ceramic "pebbles" that protect their surfaces are dislodged or worn away. Direct sunlight then hardens the asphalt, which eventually cracks and splits, allowing water to slip through. Then it's time to find a contractor.

WINDOWS

Glass, of course, survives until someone breaks it. Well-made windows can last as long as a house itself, needing only periodic painting of wooden frames. Aluminum-framed storm windows now carry warranties of up to 40 years—the expected life of their paint job. Plexiglas safety windows, required in storm doors under some building codes, seldom break, but the scratchable surface may eventually turn milky as the door is prodded open by package-carrying shoppers and the family dog. Double-glazed windows—those in which two panes separated by an air space are used to prevent heat loss—can last 30 years or more. In poorly made ones, however, the sealant can deteriorate in 2 or 3 years, allowing water to leak in and destroying the window's insulating power.

INSULATION

In a properly designed house about the only hazard to insulation is fire. Though fiberglass and rock wool themselves are not flammable, the paper backing that supports them sometimes is. Plastic foam insulations do not burn either, but high heat melts them away almost instantly. Insulation should, therefore, last as long as the house itself. The most common exception occurs when faulty construction allows water to collect inside the wall, destroying insulating power immediately. The insulation must then be replaced.

HEATING SYSTEMS

With a few exceptions, furnaces and ducting are designed to last 25 to 30 years, although major repairs will in all likelihood be required after about 10 years. The average house should reach the end of its useful life just about the same time its second furnace does.

ELECTRICAL WIRING

It isn't the wire in electrical wiring that wears out. Copper and aluminum will hold up well almost forever under normal service conditions. But the hard-rubber shielding used until about 30 years ago eventually hardened and cracked off—usually when someone tugged on it while replacing a lighting fixture— so that sooner or later enough bare wire was exposed to cause an electrical short or even a fire. Hard-rubber insulation occasionally failed in as little as 5 years, but the original wiring in some 60-year-old houses is still perfectly usable. Though 30 years was probably about average, a recent study of electrical fires in Britain concluded that the variation was so great that no design life could be predicted.

Modern plastic insulation seems to age more slowly than natural rubber. Again, however, plasticizers (the chemicals that keep plastics soft and pliable) slowly evaporate from the material, leaving it hard and brittle. The consensus is that wiring being installed today should last 40 years or so. The Consumer Products Safety Commission, however, warns that thermal insulation designed to reduce heat losses and fuel bills also reduces dissipation of heat from wiring. At best, this causes faster deterioration. When the wires are being used at or near their rated current-carrying capacity, the temperature of the wires can climb well over the 140° F allowed by safety codes. Several fires have been attributed to this cause.

PLUMBING

Both cast iron and galvanized steel are susceptible to corrosion, and water that is unusually acidic or basic can make short work of them. Though old-fashioned cast-iron waste lines have been known to last as long as the building they were installed in, water lines are sometimes eaten away in only 2 or 3 years. In most cases, a service life of 15 to 20 years can be expected. In the United States, cast iron and galvanized steel have been largely replaced by longer-lived substances, but in Europe they are still the standard materials.

Brass and copper are more durable. Brass water lines survive a minimum of 30 years, but are so expensive that they are seldom seen. Because pure copper is too soft for most applications, it is almost always alloyed with traces of other metals. Its service life varies widely with the exact composition used; 20 to 25 years is about average.

Water pipes made of plastic, notably polyvinyl chloride (PVC), have been in use only 20 years or so. Failures are rare so far, except where plastics have been melted by fire or eaten away by industrial discharges. Warm water can leach plasticizers out of the PVC, however, so plastics are seldom used for water intake lines.

Water heaters are good for 20 to 25 years of service. After that, their steel holding tanks usually become too rusty to use. Fiberglass liners used in most heaters being built today extend the service life by 5 to 10 years.

Roadways

BRIDGES

Man began building bridges as early as 2650 B.C., and these ancient structures were meant to last: The oldest existing bridge that the *Guinness Book of World Records* has been able to date is the slab-stone single arch spanning the Meles River in Turkey, constructed around 850 B.C. Today, bridges are larger, but they don't necessarily have longer life-spans. Nor are they meant to—engineers are instructed that the bridges they build must endure a minimum of 50 years. In reality they last much longer. The Brooklyn Bridge, for example, the largest suspension bridge of its day, was completed in 1883 and still remains in service.

Natural bridges, such as the Landscape Arch in Canyonlands National Park, have the longest life-spans, naturally. But eventually even these wonders succumb: In some sections the Landscape Arch has narrowed to a mere 6 feet, the victim of erosion.

CANALS

The Assyrians, Egyptians, Phoenicians, and Sumerians all built elaborate canal systems more than 2,000 years ago. As early as the seventeenth century B.C., King Sennacherib of Assyria had a 50-mile-long irrigation canal built from Bavian to Nineveh. Probably the earliest canal still in use, however, is the Fosse Dyke connecting the Witham and Trent rivers in England, built by the Romans in the first century A.D.

The death of a canal is no more natural than its birth. Canals do not wear out; they are abandoned when the economic environment changes enough to destroy their reason for being. France, Germany, Holland, Belgium, and England are crisscrossed by many thousands of miles of operating canals, many built as early as the twelfth and thirteenth centuries and nearly all completed before 1800. By contrast, the famed Erie Canal, which lined the Hudson River at Albany with Lake Erie at

Buffalo, was completed as late as 1825; yet today little remains of it but a few narrow ponds and long grassy hollows. Railroads, able to carry more cargo faster and to more destinations, made the Erie and many of its fellows obsolete almost before the canals were completed.

This is not to suggest that modern economics has made canals passé in all technologically progressive regions. Of the world's major ship canals, only four are more than a century old. The two longest—the Saint Lawrence Seaway, whose three canals connect the Great Lakes with the Atlantic Ocean, nearly 2,400 miles away, and the Volga–Don, which runs 225 miles to complete the shipping lane from the Baltic to the Black Sea—were completed in 1959 and 1964, respectively. Nevertheless, many of the world's busiest canals have already reached venerable old age. The accompanying table gives the ages of the dozen oldest major ship canals.

Canal and location	Year opened
Terneuzen–Ghent, Belgium to Netherlands	1827
Suez, Egypt	1869
North Sea, Netherlands	1876
Corinth, Greece	1893
Manchester Ship, England	1894
Kiel (Nord–Ostsee), West Germany	1895
Sault Sainte Marie (Canadian), Lake Huron to Lake Superior	1895
Chicago Sanitary and Ship, Lake Michigan to Des Plaines River	1900
Cape Cod, Buzzards Bay to Cape Cod Bay	1914
Houston Ship, Houston to Gulf of Mexico	1914
Panama, Atlantic Ocean to Pacific Ocean, Panama	1914

RAILROAD TIES AND RAILS

Conrail says that its ties and rails last about 30 years, on the average. On a straightaway, rail life can run as high as 40 years; rails situated on curves succumb to wear far more quickly. A rail wears almost solely on the inside edge, where the flange of railroad-car wheels scrapes past; rails are therefore switched side-for-side after 15 to 20 years of use, effectively doubling their service lives.

The creosoted wood ties used in this country last about 40 years. Concrete ties seen in Europe survive longer, but would be markedly more expensive in a region as heavily forested as the United States.

HIGHWAYS

The life-span of a road or highway depends on where it is located, what the weather conditions are, and how much heavy traffic it must carry, as well as on the materials used in construction. The U.S. Code, Title 23, Section 109, says that highways should be built to last at least 20 years before they need major overhauls. But the recent controversy surrounding the pathetic condition of Interstate 80 (the transcontinental route) indicates that even this seemingly short span is sometimes difficult to reach.

In New York City, where roads and highways suffer more punishment than almost anywhere else in the world, life-spans are projected according to the road surface used. If a highway is resurfaced with 1.5 inches of asphalt, the city rates its life-span at 10 years. If the highway is surfaced with 3 inches of asphalt on a 6-inch concrete base, the city estimates it will last 20 years. Anyone who has driven a car in New York City and has successfully avoided disappearance into a pothole will recognize immediately how absurd these estimates are.

Between 1958 and 1960, six road loops were built in Ottawa, Illinois, by the Federal Highway Administration to evaluate "the dynamic effect of moving vehicles on roads and bridges." At a cost of $25 million, nearly 5 miles of roadway were constructed, with variations in their composition (asphalt or concrete) and in the thickness of both pavement and base every several hundred feet. Then heavily loaded army trucks were sent around the loops until the pavement began to fall apart.

The results of the tests were somewhat vague. Although they supported the design criteria that had been devised for highways up to that point, the tests also revealed that there were so many variables—in soil conditions, construction techniques, materials used, environment, and the weight the pavement had to bear—that no firm standards of construction could be formulated that would hold for the entire country. Recently, there has been talk of updating standards to take into account the heavier loads and more sophisticated methods of construction that exist today. But the cost would be so prohibitive that there is not a legislative or administrative body in the land that will even touch the project.

It should be noted that asphalt and concrete are by no means the only enduring surfaces for roads, a point clearly demonstrated at the Indianapolis Speedway. For tradition's sake, two-fifths of the original brick surface, built in 1909, remains at the start–finish line.

TRAFFIC LIGHTS

The Department of Transportation estimates that traffic lights last from 15 to 20 years. Unfortunately, the department cannot provide an exact figure, since changing safety regulations have forced modification and replacement of the lights every 5 to 7 years. The bulbs in traffic lights are rated to last 8,000 hours and are changed approximately once a year.

STREET LIGHTS

Street-light fixtures and poles are made to last 15 to 20 years, barring chance meetings with wayward automobiles and trucks. The old mercury lamps that were the standard lights 10 years ago have been gradually replaced by new high-pressure sodium lamps. The effective life-span of the two lights is the same—about 16,000 hours or 4 years (used only at night). But the sodium lamps are more efficient users of energy.

HIGHWAY GUARD RAILS

Guard rails and concrete center medians are designed to last as long as the roads they adjoin—about 20 years—an optimistic estimate, considering the frequency of automobile accidents.

ROAD SIGNS

Signs with reflective sheeting last from 7 to 10 years, depending upon atmospheric conditions. Reflective signs facing south have significantly shorter life-spans than those facing north because of the deleterious effects of direct sunlight. Signs made of porcelain-baked enamel, aluminum, steel, and marine plywood—which depend upon prismatic buttons to reflect light at night—last 15 to 20 years.

One of the major causes of shortened life-spans for signs is vandalism. It is estimated that 10 percent of all signs erected are either stolen or mutilated before their projected life-spans have been reached—a factor thought to contribute to the rising rates of traffic fatalities in some states.

THE WHITE LINE IN THE MIDDLE OF THE ROAD

In more and more parts of the United States the white line in the middle of the road is no longer painted; instead, the road surface is coated with thermoplastic, applied at 400° F. For a high-volume highway in the eastern United States—the Long Island Expressway, for example—the painted lines in the middle of the road last only about 3 to 4 months; thermoplastic lines, on the other hand, should last up to 3 years.

Vehicles

AUTOMOBILES

The life-span of an automobile is a difficult measure to produce. For every vehicle still chugging away after its first 100,000 miles, there seems to be a comparable lemon that died before its warranty expired.

The first workable gasoline-powered automobile made its maiden voyage in Mannheim, Germany, in 1885, driven by its builder, Karl Benz. Benz constructed two such "motorwagons" that year; one of them is still in running order at the Deutsches Museum in Munich.

The life-span of an automobile obviously depends on the way in which the vehicle is maintained. Automotive writer Boyd Taylor of Atlanta, Georgia, can support this idea better than anyone else. He managed to keep a two-door 1936 Ford running for 11 complete turns of the odometer—more than 1 million miles.

While determining the average life-span of an automobile itself is an impossible endeavor, the car's various parts have fairly well accepted average life-spans. The accompanying table should provide a reasonable estimate of what to expect from various automotive parts. In most cases, however, mileage figures are more useful than life-spans expressed in years.

Automotive part	Life-span (miles, except as noted)
Air filter	10,000
Alternator	2–6 years (lasts longer on older model cars)
Automatic transmission	50,000–70,000
Battery	30,000–50,000
Brakes	
Disc-caliper assembly	60,000–80,000
Disc-pad assembly	30,000–45,000
Brake drum	100,000

Automotive part	Life-span (miles, except as noted)
Carburetor	
Domestic	50,000–75,000
Imported	30,000–60,000
Clutch disc	50,000–60,000
Connecting rods	100,000
Distributor	3–5 years
Drum lining	20,000–25,000
Exhaust pipe	3–4 years
Front-wheel bearings	100,000
Fuel pump	50,000–60,000
Manual transmission	60,000–100,000
Muffler	2–3 years
Oil filter	2,000–6,000
Oil pump	60,000–70,000
Pistons and piston rings	100,000
Points	12,000–20,000
Shock absorbers	12,000–20,000 (lifetime, heavy-duty shocks are available and worth the additional cost)
Spark plugs	10,000–20,000
Tailpipe	2–3 years
Valves	100,000
Water pump	2–5 years
Wheel cylinder	20,000–25,000

Tires. Mileage warranties today run from 10,000 to 40,000 miles. Despite a few highly publicized cases in which faulty designs and manufacturing defects have caused widespread failures of some models, most tires last out their warranties unless their owner is given to jackrabbit starts or forgets to have his wheels aligned. Failure to rotate your tires or improper inflation also shaves years off their life-spans (overinflation can cut their life-spans in half). In a few cases, high-quality radial tires have approached 60,000 miles before wearing out.

BUSES

New York City's buses, subjected perhaps to the hardest use, average about 15 years of service, going through several major overhauls in that time. Life-spans of 20 years or more are common in regions where buses see less arduous duty. The electric buses powered by overhead cables used in many cities often withstand up to 25 years of service.

As a species, buses have been around since 1662, when the French philosopher Blaise Pascal introduced a horse-drawn omnibus to Paris.

SUBWAYS

There are 67 underground railway systems in use today, the oldest being the London Transport Executive, inaugurated in 1863. The first subway in the United States was opened in Boston in 1898. The New York City Transit System, with more than 200 miles of lines, is by far the world's largest; it was opened in 1904.

New York's pre–World War II subway cars turned 40 years old before they were scrapped.

RAILROAD CARS

The average Amtrak passenger car has a life-span of 25 years, although exceptional cars last longer (one 40-year-old lounge car is still in service). Amtrak adapts its older cars to accept more modern power sources, increasing their life-spans by about 10 years. On other lines, passenger cars sometimes remain in service significantly longer. When Conrail took over commuter service in the New York City area from failing private ownerships and replaced old cars with new equipment, some cars had been in daily use for 45 years.

Most of the freight cars now in use date only from the early 1960s, when 100-ton cars and specialized freight carriers replaced the 70-ton boxcars that had been the industry standard. Life expectancies for the new cars are about 25 years—a little less for cars carrying such abrasive and corrosive cargoes as high-sulfur coal. Freight-car wheels have to be replaced every 200,000 miles or so.

Locomotives used in long-distance hauling have averaged 14 or 15 years in service for the last few generations of cars. Old engines are usually retired for economic reasons, not because they are worn out. At about 15-year intervals, new locomotives have been designed, each able to haul roughly twice as much as the models formerly available. They have been adopted rapidly to cut the extra personnel costs involved in running two trains where one could do the job.

Switch locomotives, used to shift cars from siding to siding in railyards, are used until worn out, after about 25 years.

AIRCRAFT

Airplanes. The life-span of an airplane depends both upon how airworthy it is and upon how expensive it is to operate. While commercial jets can remain in service for 20 years or more, the airlines find that jets are no longer economical after 12 to 14 years; after that period has elapsed the jets are sold, usually to foreign countries. Boeing 707s, the standard jets of our time, have effective life-spans of about 60,000 hours in flight. Other types of planes can last a lot longer: One DC-3, built in 1935, for example, is still being used for commercial service.

Helicopters. A helicopter's rotor system, which takes the most punishment, is built to function a minimum of 2,400 hours, but can last indefinitely if it is carefully maintained. Helicopters themselves have long life-spans; many of those built in the early 1950s are still in service.

Blimps. A certain tire company, which owns and operates three blimps in the United States and one in Europe, reports that the fabric coating the blimp lasts from 8 to 12 years. The gondolas (the passenger cabins hinged below the air bag) on the American blimps were constructed almost 40 years ago, in 1942, proving that, for blimps, 1942 was indeed a good year.

Gliders. The National Soaring Society estimates that most gliders, if properly maintained, last 10 to 25 years.

SHIPS

From a financial standpoint, the life-span of a commercial ship—the length of time that its yearly depreciation is still profitable for its owner—is 25 to 30 years in the United States and 10 to 15 years elsewhere. After that time, owners usually sell their vessels to lesser shipping lines or small countries. But the actual life-span of a commercial vessel is quite long: Some ships now plying the Great Lakes have been around for 50 years, and tankers from World War II still cross the oceans. Ships that sail in fresh waters last longer than those subjected to the corrosive effects of salt water.

Yachts. Some yachts still in use today date back to the 1880s. Most of them are in commercial use—as cruise boats, for example—because commercial owners usually keep their vessels in better repair than do private owners. Modern yachts, usually constructed from fiberglass, last a minimum of 25 years— perhaps longer, as only time will tell. Small pleasure craft generally last 25 to 35 years. Offshore racers, however, have considerably shorter life-spans. Naval architects are continually designing faster boats, so unless a handicapping system is used for a particular race, a vessel becomes obsolete in just a few years.

Modern Inventions

NUCLEAR REACTORS

The Nuclear Regulatory Commission licenses fission power stations for 40 years. Most of the reactor facility could probably remain in use significantly longer than that, but at the end of four decades key parts of the core and cooling equipment will have grown unreliable owing to mechanical wear and radiation bombardment. Renovation is out of the question: The reactor core and ancillary mechanisms are far too "hot" to be approached for repair or replacement. At least 12 major power reactors are scheduled for retirement by the year 2000.

Unfortunately, it seems that today's nuclear plants will be with us long after their useful lives have come to an end. Though nuclear wastes have gained far greater publicity, it is the reactor itself whose disposal presents the biggest problems. Bathed by a constant spray of neutrons throughout its service, the stainless-steel core becomes virtually as radioactive as the nuclear fuel itself. Some of the isotopes produced are exceptionally long-lived. Niobium 94, a common component of radiosteel, has a half-life of 20,000 years. For nickel 59, the half-life is 80,000 years. What this means is that it could take up to 500,000 years for the radioactivity of a commercial reactor to decay to safe levels—roughly 100 times the duration of recorded history. A few physicists claim it could take three times that long.

Roughly 70 nuclear reactors have been decommissioned so far, but many critics doubt that the experience gained will be of much help when the time comes to retire commercial reactors in use today. Most of those taken out of service were small research reactors; the largest was an early commercial power plant of barely 200 megawatts capacity. In contrast, many of the power reactors now operating are in the 1,000-plus megawatt range, and even larger ones are projected. The extra size alone is enough to complicate dramatically the problems of retiring them. For example, the stainless-steel walls of a full-sized reactor core are up to 1 foot thick. The remote-controlled cutting torches needed to take one apart have never been developed.

Once nuclear fuels, wastes, and radioactive cooling liquids have been removed from a reactor, there are three basic approaches to the task of retiring the power plant itself: mothballing, in which the reactor is simply monitored and guarded—for 500,000 years!—to make sure it stays intact; entombment, in which it is buried in concrete and monitored;

and dismantling and shipping to a storage site. Though a few reactors have been entombed and small ones have been dismantled, the most popular approach planned for large power units is a combination of mothballing and dismantling. In this scheme, the reactor would be guarded for roughly a century—long enough, it is hoped, for radioactivity to decay to levels where the core can be cut up by shielded workmen instead of by costly remote-control equipment—and then taken apart.

Assuming that this arrangement works as planned—and few of nuclear power's critics are willing to make such an assumption—a key problem will remain. No one knows where radioactive core components can be safely hidden for 5,000 centuries. At the current rate of technological progress, large-scale decommissionings will be upon us long before any final answer is found.

COMPUTERS

The original computers, dependent upon mechanical relays, underwent constant wear and tear and often broke down. Since the introduction of integrated circuits, however, the number of computer parts that are vulnerable to the aging process has decreased dramatically. The tiny electronic chips that now function as a computer's brain either fail quickly or do not fail at all. As one computer technician said: "If it lasts a moment, it'll last an hour; if it lasts an hour, it'll last a day; if it lasts a day, it'll last a month; if it lasts a month, it'll last a year; and if it lasts a year, it'll go on forever."

The reason for a computer's durability lies in the nature of the tiny silicone chips that make it work. If impurities or defects mar the integrity of a chip, the malfunction will surface as soon as the computer is put into service. If there are no impurities, the chips, which do not have to endure any physical wear, can last indefinitely. Thus the life-span of a computer is determined by how soon it becomes technologically obsolete, not by how soon its parts wear out.

TELEPHONES

Ma Bell's equipment is all designed to last about 25 years, according to the engineers at Western Electric. After that, the cost of keeping a telephone, transformer, switching unit, or whatever in working condition makes it more economical to take these devices out of service even if they are still functioning. During its life, a telephone will have gone through several dials or sets of push buttons and have had many other parts adjusted or replaced.

The first transatlantic telephone cable, TAT-1, was laid in the mid-1950s and retired after 20 years of constant use. The cable carried only 50 messages at a time, the same amount of power required to send 4,000 calls over a modern line. It was shut down because it was too expensive to use, not because it had become unreliable. TAT-1 had been accidentally broken by fishing trawlers several times, but it had never failed on its own.

TELEVISIONS

A TV set has two weak spots: the picture tube and the mechanical tuner. Even in televisions claimed to be constructed entirely of solid-state components, the picture tube contains a wire filament that will eventually burn out, just as that of a light bulb does. Tubes usually carry 2-year warranties, but most survive at least 3 years; 6 years of normal use is about as long as you can hope for. Mechanical tuners wear out because friction abrades metal contacts whenever the channel is changed. They are likely to need repair within 5 years. Given a second picture tube and a second tuner, a good television should survive about 10 years; a few make it to age 15.

STEREO EQUIPMENT

With their life expectancies rated at 20 years or more, loudspeakers are the most durable element of any stereo system.

Short of violent impact, about the only thing that will damage them is having the volume turned up so high that it overstresses the speaker cone and driving mechanism. Apartment dwellers are unlikely ever to damage a loudspeaker.

Purely electronic components—tuners, amplifiers, and the like—have life-spans of 10 to 15 years. In theory, transistors themselves should last indefinitely. What eventually happens to electronic gear is that a resistor or capacitor burns out or the insulation on a wire breaks down, causing a short.

Turntables, tape decks, and other electromechanical components are subject to friction as well as electronic breakdown and have the shortest lives of any stereo equipment. Still, the best turntables should last 10 to 15 years. Lower-quality equipment is likely to give out after 5 or 10 years.

To a great extent, the life of stereo equipment depends on its use at the hands of the owner. Components may be designed to survive a decade of home use, but manufacturers assume the equipment will be on only 2 or 3 hours a day. Full-time listeners may well find their stereos giving out halfway through their expected lives. Professional equipment is designed to last through a decade of 18-hour working days, but the cheapest studio turntables and amplifiers cost more than twice as much as top-quality home models.

Stereo consoles, with turntable, tuner, tape deck, and often speakers all built into a single unit, don't seem to last as long as component outfits, perhaps because some of the money that would otherwise pay for electronic parts goes into cabinetry instead. Oddly enough, this seems to make little difference in the time consoles spend in the purchasers' homes. One industry economist suggests that stereo consoles are more furniture than stereo unit; they are bought for appearance and fulfill their true function even after the turntable or tape deck has failed.

Nearly all phonograph needles now sold are made from industrial diamonds and will withstand about 1,000 hours of use before they need to be replaced. Elliptical, shibata (a needle that is flatter along its sides), and other specially shaped needles designed for highest-quality sound reproduction last slightly longer. The few sapphire needles still in use should be replaced every 40 to 50 hours.

PHONOGRAPH RECORDS

All records are made of chemical components that evaporate over time, so whether you realize it or not your record collection is dissipating into chemical fumes right before your eyes—albeit very slowly. What really takes a toll of records, however, is improper handling—leaving them sitting out in the air where dust and other dirt particles can settle on them, laying them on top of a radiator or other source of heat so they warp, or stacking them up three and four deep on the turntable spindle where they may grind against each other. Ordinary precautions, such as handling them by the edges, playing them one at a time, and using a lightweight pickup will help LPs age gracefully and well. Moreover, records are manufactured better today than they were in the past.

In general, the well-cared-for record can last 20 to 30 playable years. You can extend this life by doing what some audiophiles do: Tape the record when it is fresh out of the jacket and play the tape recording, not the record, in the future (to find out how long that will last, look under "Recording Tapes").

If a record doesn't sell, it may come back to haunt you by being recycled into another record. Unsold records are ground to powder, melted down, and made into new records. At one time the entire record, label and all, was used, but manufacturers found that tiny shreds of paper from the label got trapped in the plastic, absorbed moisture, and in time caused the new record to warp. Now they avoid this problem by punching out the center portion before powdering the discarded record. So that album of 20 great Estonian drinking songs you passed by in the record shop a year ago may be reincarnated as part of the album of Bach cantatas you bought today.

RECORDING TAPE

Whether it's packaged as a cassette or on an open reel, sound-recording tape, manufacturers say, has an indefinite life-span. In fact, at least one manufacturer, TDK, guarantees its cassettes for the buyer's life, at normal use. (It has made good on its guarantee even when deceased buyers' descendants have sent for replacements of cassettes that had outlived their owners.) The same applies to recording tape stored on an open reel. There is at least one instance on record of a reel of tape that was played for 86,549 hours providing background music almost continuously over a period of 10 years with no tape failure. The average tape does not get this kind of brutal workout. Experts estimate that in a year a cassette, for example, is played about 100 times—200 passes (play and rewind) through the tape machine. With this kind of use sound quality does deteriorate somewhat after 2,000 passes, about 10 years, but not significantly. Tapes have been played the equivalent of 100 years in laboratory tests and survived intact.

For maximum durability and tape use, keep tapes in a cool, dry place. Leaving a cassette on the car dashboard or in the trunk, for example, may warp the plastic case with the result that the tape won't track accurately on the tape head. Tapes on reels should be stored like records—vertically, to prevent the reels from warping. Warped reels may scrape oxide particles, the magnetic dust that captures the sound, off the edge of the tapes and distort recordings. (Most tapes are made with a tensilized polyester base and a layer of either pure ferric oxide or ferric oxide and cobalt that arrange themselves according to magnetic signals from the taping head.) If you have a warped reel, transfer the tape to a reel that is not warped.

Tapes, unlike records, thrive on playing, because it relieves some of the tension of fast rewinding; but since all tapes do shed, you should clean the tape path every 20 plays to minimize wear and tear. Use cotton swabs and denatured alcohol on the track and on the heads themselves.

VIDEOTAPE

Recording sight as well as sound is much harder on tape; for that reason, firms selling videotape for home recording units you can hook into your television set are much more conservative about their guarantees. In most home videotape machines on the market today, the tape is moved past a highly polished drum that is itself moving, spinning at an average rate of 1,800 revolutions per minute. This takes a tremendous toll of the surface of the tape, which is wrapped halfway around the drum; hence, no tape is expected to last more than 200 passes (100 plays) in top-quality condition.

JUKEBOXES

A modern jukebox has a useful, active life of about 5 years. But proper maintenance and treatment could easily increase its life-span to 20.

Communication or Artistic Expression

PIANOS

A well-constructed and carefully maintained piano is virtually indestructible, according to John Steinway of Steinway and Sons, the most celebrated of piano makers. Steinway sets an arbitrary limit of 50 years on the pianos they will repair or rebuild, but only because of the enormous cost of fashioning outmoded replacement parts by hand.

The best way to maintain a piano is to play it. A piano that is used only to support a lamp and framed photographs will slowly deteriorate. Pianos should be tuned at least four times a year, even if they are not played. They should also be kept at constant temperature and humidity, away from direct sources of heat. Steam heat is especially damaging to a piano: If the instrument is kept in a steam-heated room, a humidifier or at least a forest of green plants should be installed to provide moisture.

There is no such thing as a vintage piano. In fact, thanks to such things as improved glues, new "action" designs, and such space-age materials as Teflon, recently constructed pianos are thought to be considerably superior to older models. The most noticeable change in pianos built in the last 20 years is the use of plastic rather than ivory keys. Says John Steinway: "We shot the last elephant in 1957." Plastic, while perhaps not as esthetically pleasing as ivory, is a superior keyboard material. Ivory is naturally yellow and must be bleached, and it is easily chipped. Furthermore, an elephant tusk is not large enough for a key to be made from a single slice of ivory. Thus an ivory key had to be made from two pieces of ivory with a seam or joint.

Probably the most durable of all pianos was made by Steinway and Sons during World War II for use by the armed forces. Painted army green or battleship gray, it was called the "GI piano" and was sturdily built—with nails as well as glue, for example—to withstand the rigors of wartime. There were 3,500 of these pianos made, and they can still be found on military installations and in the homes of retired officers.

VIOLINS, VIOLAS, AND CELLOS

Stored in a museum case with dust excluded and humidity controlled, a violin, viola, or cello could last virtually forever. A few have made a good start on immortality even without extreme precautions. The first violins appeared early in the sixteenth century. Instruments from that period are rare, but they do exist. At least three cellos remain from the work of Andrea Amati, first of the great instrument makers of Cremona, Italy, who died around 1578. The famed Antonio Stradivari, probably greatest of the Cremona violin makers, worked from 1658 to 1737, producing at least 1116 instruments. Of these, nearly 600 still exist, including 540 violins, 12 violas, and 50 cellos. The oldest dates from 1666.

It is not only their rarity that makes early stringed instruments sought after today. As fine violins, cellos, and their relatives grow older, their tones grow fuller and more brilliant. Exactly why is unknown, though many experts believe that the wood itself grows more flexible with age and, therefore, is better able to vibrate freely when the strings are bowed. Whatever the explanation, nearly all the truly great violinists and cellists performing today use instruments made at least two centuries ago. Yehudi Menuhin, for example, has used a violin made by Giuseppe Guarneri in the first half of the eighteenth century. The Tokyo String Quartet plays instruments by Nicolo Amati, grandson of Andrea, on semipermanent loan from Washington's Corcoran Gallery.

Stringed instruments need surprisingly little care to remain in good condition: They should be wiped free of rosin after each use to keep the pores of the wood from becoming clogged. Dust inside should be removed periodically. (The traditional method is to pour dry, slightly warm, uncooked grain cereal inside and shake it around gently, but vacuum cleaning will do as well.) It is most important to keep the instruments away from drafts and to wrap them warmly if they must be taken outside in order to protect them against changes in moisture content. At best, sharp changes in temperature will ruin the instrument's tone until it adapts; the shock can split in half an instrument that would otherwise have lasted centuries.

WOODWINDS

Unlike stringed instruments, a woodwind is never better than in its first year, and its tone deteriorates rapidly. The average life of an oboe is less than 10 years; almost none lasts out the second decade, and many succumb after only 5 years. In fact, the upper joint of an oboe often cracks within the first 6 months, particularly if the instrument is played more than 1 hour a day during that period or if the hard African wood used is subjected to abrupt temperature changes. Other woodwinds have similar life expectancies, but few are as prone to cracking in their early months.

Woodwind players must cut their own reeds, the thin vanes whose vibrations generate the instrument's sound. The highest quality reeds are cut from a cane known as *Arundo donax*, which grows best in southern France, near the mouth of the Rhone. The process is extremely delicate, and many skilled professionals must discard four out of five reeds they make. At most, a reed will last 2 weeks. Some perfectionists use a reed for only one performance, reducing the reed's life-span to 2 or 3 hours.

GUITARS

They do make guitars like they used to—exactly like they used to, according to Matt Umanov, a New York City guitar dealer and repairman, who says he has taken literally thousands of guitars apart to check their construction. Though only one American guitar company still offers a lifetime warranty on its instruments, a top-quality guitar will easily last 200 years if it is treated carefully. "The problem is that most people buy them like toaster-ovens," Matt comments. "They happen to have $700 in their pockets and decide to buy the best. Then they kick the hell out of it."

The toughest guitars are electrics. An electric guitar's body is made either of solid wood or of three-ply maple; the neck can warp or twist, as it can on an acoustic guitar, but the body is virtually indestructible.

Acoustic guitars are more fragile, but only a little. A lesser instrument eventually becomes warped as the tension of the strings pulls the face of the guitar up. This raises the strings away from the neck, making the guitar difficult to play. Like other wooden instruments, guitars must be protected against abrupt changes in temperature and humidity. More than one guitarist has carelessly left his instrument in the cargo hold of a jet and found the face or back split when he recovered it.

BOOKS AND PAPER PRODUCTS

Perhaps more than anything else produced today, books seem programmed to self-destruct. In fact, a book published hundreds of years ago will frequently be in better shape than last year's best-seller. There are two major reasons why books, magazines, and other paper products do not last as long as they used to. The first is that books have now become consumer products rather than luxury items and thus are manufactured with considerably less care than they once were. Second, the paper's rag content—the amount of wood fiber as opposed to pulp found in the paper—has steadily decreased over the years. Paper with 100 percent rag content is virtually unobtainable today, and most papers range in rag content from 25 to 75 percent. The higher the rag content, the longer the paper will last. Paper designed for legal documents and financial ledgers, for example, is usually 75 percent rag, since it may have to be kept over long periods of time.

There are two serious problems with pulp paper: It dries out easily (rag in paper tends to trap moisture), and it is highly acidic and therefore becomes brittle, discolored, and apt to crumble. Paper can be deacidified, but that is an expensive process.

There is much that can be done to preserve books, even those made from today's poor quality paper. Libraries, for example, rebind books to lengthen their life-spans. Here, in a process usually called library binding, the books, often held together only by glue, are taken apart and the signatures are stitched back together. By this technique a book's effective life-span can be increased from about 25 readings to more than 100. For more serious collectors, the life-span of a book can be extended almost indefinitely by such techniques as freezing, vacuum storage, and even by coating each individual page in plastic.

Around the home the chief dangers to books and papers are humidity (too much or too little), sunlight, air pollution, and various fungi and molds. Thus, open bookshelves offer a book a shorter life-span than shelves equipped with such things as glass doors. Books with leather bindings should be occasionally treated with lubricant to protect them from drying and cracking. Cellophane tape and masking tape should never be used to repair the pages of books, for the acid contained in the tape will cause the paper to deteriorate. There are special nonacidic tapes designed for book repair. Letters and other paper documents should be kept in airtight, moisture-proof, opaque containers to assure maximum life-spans.

WRITING MATERIALS

Pen-and-Ink Fountain Pens. A high-quality fountain pen has an indefinite life-span. Well maintained, it could pass from generation to generation of regular use. Most important is proper cleaning: Fountain pens utilizing water-soluble ink can be cleaned with mild detergent and water; those filled with other than water-based inks require special solvents for cleaning.

Ball-Point Pens. The manufacturers of ball-point pens rate their products according to how long a line the particular pen will draw. The range of available ball-point pens stretches from models that will draw a line 2,500 feet long to those that will produce a line more than 15,000 feet long. Most ball-points on the market are rated from 4,000 to 7,500 feet. Some ball-points are refillable, and the available refills range from 4,000 to 10,000 feet.

Most ball-points manufactured today, however, are disposable, and most, it seems, are disposed of before they run out of ink. Says a representative of one of the larger manufacturers: "The only consumer reaction we get is when someone sends us a dry pen and exclaims, 'Isn't this remarkable? I managed to hang onto it long enough to use it up!'"

A ball-point pen does have a definite shelf life, however. Many pens with black ink should be kept only 1 year or less;

those with blue ink, only 2 years. Ball-points specially designed for use with copying machines have even shorter life-spans. Pens encased in plastic "blisters" can be stored longer than pens packed loose.

Expiration dates are not usually listed on pens or on their immediate containers. The box in which the retailer receives the pens, however, should be marked with an expiration date or a batch number.

Soft-Tipped Pens. The range of soft-tipped pens available stretches from pens that will draw a line 2,000 feet long to pens that will supposedly draw a line more than 12,000 feet long. The shelf lives of soft-tipped pens are about the same as those of ball-points, although the former may dry out more quickly if they are not sealed in moisture-proof packages. In most cases the tip of a soft-tipped pen deteriorates before the pen runs out of ink. Tips are made from felt, nylon, and plastic, plastic being the most durable.

Quill Pens. A true quill pen—a simple bird feather with no metal tip — can be used for up to 6 months of regular writing if it is properly maintained. The point can be shaped or trimmed about 12 times. Wear will be determined by the care given to the quill, the abrasiveness of the paper used (rough paper will act like sandpaper on a quill), and the quality of ink. Water-soluble inks will allow the quill a longer life-span. When not in use, the quill should be kept with its point in water or ink. If it is left in the ink bottle, add a few drops of water to the bottle daily and then flush out the bottle and add fresh ink every few weeks.

Pencils. A hard pencil can draw a line more than 30 miles long or write more than 30,000 words.

Inks. Inks generally sold for use in fountain pens can be kept for more than 1 year. Exotic inks—those cured by ultraviolet light, for example—may last as little as 3 months. Two-part inks (those mixed by the printer before use) will last less than a day. In general, the darker inks (blue and black) can be kept longer than lighter colored inks.

Once printed, black and blue inks can last indefinitely, depending on the formulation. Colored inks are considerably more fragile. Ink is formulated according to its use: Newspaper ink, for example, has to last only 1 day and so is not particularly stable. Ink designed to be used for outdoor display, on the other hand, is specially formulated to withstand such things as sunlight and moisture and will last up to 1 year.

TYPEWRITERS

The folks at Smith-Corona Marchant say they regularly test their machines and design them to reach the equivalent of a 10-year service life. There are a few hidden assumptions built into that figure, however. Typewriters are built with widely varying levels of durability. A superlight portable intended for occasional letter writing would be beaten flat by a year of daily use by a professional typist. Nonetheless, some typewriters survive for periods that are little short of amazing. The first portable typewriter, the Corona Model 3, was marketed from 1912 to 1941. SCM still makes—and sells—ribbons used only on that machine. Their best guess is that at least 100,000 of these durable little typewriters are still in existence, many of them in regular use.

TYPEWRITER RIBBONS

There are two sorts of typewriter ribbons generally available— those that are reusable and those that are not. Carbon ribbons, which can only be used one time, can type about 120,000 characters on the average. Reusable ribbons are made from cotton, nylon, and sometimes silk. Their life-spans, considerably more than 120,000 characters, are determined by the machine on which they are used and the fastidiousness of the user.

CARBON PAPER

Carbon paper is produced in a variety of qualities and is designed for a wide range of specific uses. A sheet of high-quality carbon paper sold for general typing can be reused as many as 20 times. The paper re-inks itself, and its ultimate demise is more often a result of paper wear than lack of ink. If 20 duplications are not enough, there are carbon sheets made of plastic rather than paper that can be reused up to 40 times.

SKYWRITING AND SKYTYPING

Few things are more ephemeral than a message written in the air. The average skywritten word, spelled out in a snow-white stream of water vapor, lasts an average of 5 to 7 minutes before it evaporates from view. On a calm day the message may stay up longer. Skywriter Jack Strayer, who spells out PEPSI in mile-high letters, once had a word linger for as long as 54 minutes.

Skywriting is becoming a thing of the past and is being replaced by skytyping—spelling out words with carefully orchestrated dots of water vapor often emitted under computer control. Skytyped messages are usually limited to 20 to 25 words because that is the maximum number that can be put up in the sky and still be legible to a stationary observer on the ground. Skytyped messages may stay legible for as long as 20 minutes before fading from view.

CAMERAS

The durability of cameras has changed markedly since the Great Camera Boom of the early 1970s. Before, only professionals and serious photographers were willing to put up with the degree of skill and knowledge needed to operate a 35-millimeter, single-lens reflex camera correctly. Since then, however, automation and various accessories have made the single-lens reflex accessible to practically everyone. At the same time, the high quality required by the professionals has succumbed to the lesser demands of the mass market.

As Marty Forscher, the proprietor of Professional Camera Repair in New York City, says: "The trend in camera design today is away from the Rolls-Royce concept and toward that of the Chevrolet—with the exception that the Chevy has an engine which might last 120,000 miles even if the body falls off. I don't see even that kind of quality in many of the newer cameras.

"The rugged, well-built cameras, like the Nikon F and the Leica M series, were the hallmarks of professional photographers," Marty continues. "Used every day, and taking thousands of pictures a year in wide-ranging, sometimes hostile conditions, those cameras had a life expectancy of 10 to 12 years. They were so tough we called them 'hockey pucks.'

"In 1970, this started to change. Cameras now offer sophisticated features—auto-wind, no-focus, automatic exposure—at comparatively low prices, but these are Mickey Mouse mechanisms without lasting qualities; they are not even built of top-notch materials, and are no use to a professional who depends on his camera for his living. A professional with a new-generation camera, working it as hard as he worked his old cameras, can expect a maximum of 3 years from it."

When a modern camera starts to fail, its automatic rewind mechanism, which is usually not constructed as well as the mechanisms of the old manual winders, succumbs first. Shutters may stick open or closed; reflex mirrors may refuse to return to position; and light leaks may develop as lens mounts loosen or as hinged backs become warped.

Cameras used exclusively by amateur photographers—Kodak Instamatics, inexpensive Polaroids, and the like—have life-spans that are rated not on the basis of the durability of the camera but on the average length of time it is used after it is bought. Manufacturers estimate that these cameras are kept for about 6.5 years before they are either discarded, handed down, or traded in for another model equipped with some newer breakthrough.

PHOTOGRAPHIC FILM, PRINTS, AND PAPERS

Film. Any film—whether color or black and white—can last indefinitely when stored in a freezer. Black-and-white film will also record acceptable pictures after long periods of storage in the refrigerator. But color film, even in the refrigerator, will go stale, although its color balance will not be affected. Any film that is stored in a hot environment, in an environment that is subjected to certain types of radiation, or in a container that is not utterly light-proof will be worthless long before the "develop before ..." date stamped on its package has been reached. That date tells you when the film is 1.5 to 2 years old. There are, in addition, films used for highly specialized purposes that have less stable emulsions. These films have far shorter life-spans than the commercially marketed films.

Color prints, no matter how they are stored, eventually fade and lose their color. Scientists now place the life-spans of color prints that are carefully stored at about 50 years, making them useless for such historical records as time capsules. The information in a color photograph can be preserved, however, in two ways. First, the color print can be broken down into its primary color components and preserved as three black-and-white slides—each representing one of the color print's primary colors. Second, the color at each point in the photograph can be translated into a number or code, and this numerical information can be stored. The numbers can then be decoded and the photograph reconstructed at a later date by a kind of paint-by-numbers system.

Black-and-white prints, if they are kept at low temperatures and are not exposed to excessive light or environmental pollutants, have life-spans of hundreds of years—or so it is believed. The first photographic negative was made in 1816 by Joseph Nicéphore Niepce. This first photograph proved unstable, but a negative Niepce produced in 1826 remains legible.

Photographic Papers. Color photographic papers are never dated, mainly because they are so unstable that the manufacturer is leery of guaranteeing them for even a day of storage in retail outlets and homes. If they are kept in the freezer, the emulsion on the papers' surfaces becomes inert, and the papers themselves will then last indefinitely. Kept in the refrigerator, they will last about 6 months.

Black-and-white photographic papers, on the other hand, are extraordinarily stable. Kodak puts a 2-year date on them; Agfa, a 5-year date. In the freezer, they will last forever; in the refrigerator, at least 5 years.

Darkroom Chemicals. A variety of developers, fixers, and washers is used for different types of film and processes, and the shelf lives of these products vary significantly. These are their approximate life-spans:

Type of Chemical	Full Bottle	Half Bottle	Tray	Gallon Tank
Developer for film, plate	6 months	2 months	24 hours	1 month
Developer for paper	4–6 months	1-2 months	24 hours	
Stop bath	Indefinitely		3 days	1 month
Fixing bath	2 months		1 week	1 month
Washing aids	3 months		24 hours	

PAINTINGS

From the earliest times that man began creating works of art until the eighteenth century, painting was an art form dedicated to posterity. The guilds (ancient unions) and workshops of the Middle Ages were founded upon the highest principles of quality; their members' craftsmanship and technical skill had been passed down and refined by generation after generation. Partly because the master painters were artisans as well as artists, partly because the consumers demanded and understood quality, paintings were rendered to last. Today, paintings over 500 years old retain their original freshness and have about them an aura of timelessness; if properly cared for and restored, they may well last forever.

In the last 150 years, however, the goals of art have changed, as have the raw materials. Industrialization and commercial preparation of the materials used have brought about a decline in their quality; the disappearance first of guilds and workshops, then of royal patronage ended the ability of master craftsmen, artists, or connoisseurs to control both the nature of what was painted and the quality of workmanship; success and fame—watchwords that seemed to hold meaning only within the span of one lifetime—became all important. As a result, painters began to worry more about productivity than about durability, about how to paint the most in the shortest time rather than about how to preserve what had already been rendered. Consequently, paintings today seldom have the capacity to survive more than a few decades or centuries after their completion.

Despite the quality and durability with which the old masters imbued their work, it is likely that many of them would be hard put to recognize their own works today without a thorough education in art history. Paintings may survive for millennia, but they are constantly mellowed by time. Some colors fade; others grow darker; still others change completely, shifting relationships within a work and changing its tonal quality. A painting of the Virgin and Child by Raphael, hanging in the Metropolitan Museum of Art in New York City, for example, now depicts the Virgin, and several shepherds gathered around her, as wearing midnight-black robes, dusted with yellow sparks. Raphael had originally painted those robes a light blue, but the intervening centuries have darkened both their color and their symbolic significance.

But even the masters of the Middle Ages, with their techniques and quality, are mere dilettantes compared with other painters in human history. Paintings of the great civilizations of Rome, Greece, Egypt, and China still survive; and the cave paintings in Lascaux, France, are estimated to be at least 15,000 years old.

Around the House

MAJOR APPLIANCES

Life-spans can be very tricky things to compute when it comes to the major appliances we use today. Both government and private testers have devised different yardsticks to get some realistic notion of the longest practical lifetimes of appliances. For example, engineers at MIT devised a system that measures a product's *life-cycle cost*. By their standards a product becomes uneconomical or financially "dead" once the costs of maintaining and repairing it exceed its original purchase cost. By these standards a refrigerator's life-span is 14 years and a color TV's about 10. Of course, if you happen to get a lemon, one that is always in the repair shop, its life-span will be appreciably shorter.

A little less theoretical and more pragmatic is a measure called *service-life expectancy*, long used by researchers with the U.S. Department of Agriculture to gauge how long certain appliances last. Although some interpret the phrase to describe the entire operating lifetime of an appliance from the first day it is plugged in to its last mechanical gasp, the USDA researchers simply use it to mean the use expectancy under one-family ownership.

In a survey published in March 1978, USDA researchers Katherine S. Tippett, Frances M. Magrabi, and B. C. Gray came up with the following service-life expectancies for several common household appliances, calculated for both brand-new and used (one prior owner) examples.

Appliance	Service life (years)	
	New machine	Used machine
Ranges		
Electric	12	6
Gas	13	7
Refrigerators	15	7
Freezers	20	9.3
Washing machines	11	5
Dryers		
Electric	14	5
Gas	13	5
Dishwashers	11	7
TVs		
Black and white	11	5
Color	12	5

Other government studies for particular appliances have also been made public. For example, the Office of Technology Assessment, given the task of finding out what brands of air conditioners would give the most service for the lowest overall cost when used by the armed forces, found that air conditioners last about 10 years under normal conditions. Constant use in hot, dusty climates cuts the average life-span in half.

Yet another approach to determining product life-spans has been taken by the Energy Research and Development Administration. In assigning an average useful life to various pieces of equipment, they calculated not only the amount of time it is feasible to keep and operate a piece of equipment before it starts to wear out but also at what point the machine becomes an energy and money waster. Among the more common pieces of equipment and their average useful lives are:

Equipment	Average useful life (years)
Boilers	20–25
Electric furnaces	10
Electric heating, add-on	10
Electric motors	20–25

Fans	20
Gas-fired furnaces	10
Heat pumps	10
Oil-fired furnaces	10
Room air conditioners	8

SMALL APPLIANCES

Precise information on the life-spans of such products as blenders, toasters, mixers, and broiler-ovens is almost nonexistent, save in the laboratories of manufacturers, who have little interest in making it public. In what may be the only public test of a small appliance on record, the National Bureau of Standards recently examined several brands of hand-held hair dryers and found it impossible to predict how long any of them would last. Average service life varied from as little as 100 hours of actual use for one model to as much as 1,500 hours for another. Though all machines of a given model were supposedly identical, their durabilities varied markedly. In several cases, the longest-lived machine survived five or six times as long as its shortest-lived twin. Roughly half the dryers lasted through 200 hours of use, and the NBS engineers were left to question whether a longer design life would offer any practical benefit. Their survey found that few machines operate more than 50 hours per year; 25 was closer to average.

At UCLA, W. David Conn has surveyed nearly 2,000 households to find out how long a variety of appliances lasts. Included in the poll were toasters, toaster-ovens, can openers, blenders, coffee makers, skillets, bonnet-type hair dryers and blow dryers, electric toothbrushes, irons, vacuum cleaners, radios, and black-and-white TVs. Blow dryers, blenders, and skillets, he found, had the shortest service lives; 50 to 70 percent were used for 2 years or less. Only coffee makers, bonnet-type hair dryers, and vacuum cleaners had a median life expectancy of 7 years or more. Electric toothbrushes, AC-powered radios, and irons all had 5 to 6 years' longevity. Conn noted, however, that many of the appliances were still working when discarded or put into storage. Most owners either decided they had no use for the appliance or wanted a new model.

LIGHT BULBS

The first incandescent light bulb, tested by its inventor, Thomas A. Edison, in 1879, burned for 40 hours. Today, the average life-span of a 100-watt incandescent bulb is 750 hours and for a 25-watt bulb, 2,500 hours. Manufacturers of light bulbs can and do make incandescent bulbs that will last virtually forever. Unfortunately, there is a trade-off involved: As the life-span of the bulb increases, the bulb's efficiency (the amount of light it produces for the energy it consumes) decreases. One major bulb manufacturer, for example, has a 60-watt incandescent bulb that has a life-span of 1 million hours. The difficulty is that it takes three of these bulbs to equal the light output of a standard 60-watt bulb. The additional cost in electricity to burn three bulbs rather than one is more than $4 for every 1,000 hours of use.

Perhaps the simplest way to avoid the nuisance of changing light bulbs is to burn bulbs designed for 220-volt lines in 110-volt sockets. The bulbs will burn at only about half the listed brightness and with a yellowish light, but they will seldom, if ever, have to be replaced.

Contrary to popular belief, the number of times a light bulb is turned on and off has little to do with its life-span.

FLASHLIGHT BATTERIES

Any flashlight battery is a sealed package of chemicals that will degenerate in time. Today, you have a choice of two general types: the old standby, carbon zinc, and the longer-lasting alkaline. It's difficult to measure life-spans of batteries in use because different items, such as flashlights, transistor radios, and tape recorders, make varying energy demands. Many appliances made today have built-in shutoff circuits. For example, once a battery for a radio or tape recorder slips below the 8-volt level, the machine won't work; for all practical purposes the battery inside it is dead even though it may be capable of generating, say, 7 volts of power. Battery manufacturers avoid the confusion that could come out of attempting to devise an across-the-board measurement by instead determining a life-span measure they call service-maintenance period, better known as shelf life. By this standard, a battery is considered dead as a salable item once its chemical content has degenerated to the point where it can produce only 90 percent of its capacity. The accompanying table should give you some idea of how the power of the two main kinds of batteries slips over the years of just sitting on a shelf.

Alkaline battery		Carbon-zinc battery	
Storage time (years)	Power capacity remaining (%)	Storage time (years)	Power capacity remaining (%)
1	95	1	90
2	90	2	80–90
3	85	3	75

In general, battery experts say, you should use a battery no longer than 4 years after buying it if you want to get your money's worth.

KITCHEN KNIVES

The hardness of a knife blade is usually expressed in what is called the Rockwell scale. According to food expert James Beard, the ideal hardness for a kitchen knife is between 55 and 58 degrees Rockwell. Harder blades will not sharpen; softer ones are too easily damaged.

Until fairly recently there was an ongoing debate among

cooks about the best material from which to fashion a knife. The choices were stainless steel and carbon steel, and there was a trade-off involved in the use of each. Carbon-steel knives are easier to sharpen than those made of stainless steel and, once sharp, hold the edge longer. However, carbon steel rusts and pits easily, especially in coastal areas where moisture and salt are prevalent, and it stains when used to cut such high-acid foods as tomatoes and citrus fruit. Stainless steel, on the other hand, does not rust or stain, but it is much harder than carbon steel and thus more difficult to sharpen. (Most sharpening tools found in the home are themselves softer than stainless steel and therefore will not sharpen it.)

Today the debate has been quelled by the introduction of what is called high-carbon or no-stain stainless steel. This new material will not rust or pit, but it is also soft enough to be sharpened.

The other part of a knife that directly influences its life-span is the handle. Knife handles are usually made from wood, plastic, hard rubber, or steel. Handles made from wood impregnated with plastic are the longest lasting and will stand up to the wearing effect of a dishwasher. Wood handles can shrink, so they should not be allowed to soak. The best wooden knife handles are made from Brazilian rosewood, a hardwood with an irregular grain that resists splitting and cracking.

Again, as James Beard points out, the banner "Never Needs Sharpening" on a knife box usually means the knife cannot be sharpened and thus will have a limited life-span. All this having been said, a quality kitchen knife will last for upwards of 10 years of regular use if it is well maintained.

WAXES

Unopened, most waxes—including shoe polishes and furniture polishes—can be kept for an almost unlimited period of time.

Waxes are solid at room temperature, but can melt under heat and thus should be stored in a cool place.

So-called liquid wax is simply wax contained in a solvent, most often alcohol. Once opened, the solvent will begin to evaporate, making the wax difficult to apply. Liquid wax may cloud if it is stored for any length of time, but this will not affect its performance.

NATURAL SPONGES

A natural sponge takes about 50 years to grow to what the sponge industry calls "bath size." Once harvested, the sponge can be stored for an indefinite period. Sponges have been found in the tombs of Egyptian kings, for example, still resilient after thousands of years. Natural sponges should be rinsed thoroughly after every use, but never in extremely hot water or the fibers will tighten and the sponge will lose its resilience. A brittle sponge can frequently be restored by soaking it in washing soda.

SYNTHETIC SPONGES

Most synthetic sponges are made from cellulose that is dyed and sometimes treated with antibacterial agents. While they are not nearly as durable as natural sponges, they can be sterilized by boiling. Cellulose sponges stain and discolor easily and can be difficult to clean. If soaked in strong bleach, the sponge will disintegrate.

GARBAGE

Our trash may fascinate archeologists a few hundred years from now, but for the time being it is at best a nuisance and at worst a real health hazard. How long we must put up with it depends on what we've thrown out, as the accompanying table indicates.

Item of garbage	Life-span (years, except as noted)
Orange peels	1–24 weeks
Paper boxes	2–20
Wool socks	1–5
Plastic-coated paper	5
Polyurethane	10–20
Plastic bags	10–20
Photo film	20–30
Polyester cloth and nylon	30–40
Leather footwear	50
Plastic jars and bottles	50–80
Aluminum cans	80–100
Nuclear wastes (plutonium)	24,400 (the half-life of a given amount)

In the Tool Shed

LAWN AND YARD TOOLS

Most garden tractors, lawn mowers, and snow blowers come with manufacturers' warranties of only 1 year. Given proper care, their life expectancies are far higher. A good garden tractor should last 10 to 15 years, and many make it to age 20. The high operating speed of a rotary mower's engine cuts the average life-span to between 7 and 10 years, assuming the machine is used two to three times a week. Snow blower engines work at considerably fewer revolutions per minute and are used less frequently, so they tend to last longer than those of rotary mowers. Probably the longest-lived piece of lawn equipment is the reel-type lawn mower. Large commercial mowers run 8 hours a day, every day, all summer, and often last for decades. Unfortunately, few household yard tools survive anywhere near as long as they could. Misuse and simple neglect destroy more lawn mowers than all other causes combined.

Electric lawn mowers, hedge trimmers, and similar electrically powered devices have a wide range of service lives. Even inexpensive models should last through 2 or 3 years of regular use. More costly, heavy-duty machines will survive 10 to 15 years. What governs life expectancy is the quality of the motor and, oddly enough, the length and size of the cord (in non-battery-powered models). Long or fine-guage wire has a high resistance and cannot feed enough electricity to keep an electric motor working properly. The motor soon burns out. Most electric lawn tools can be used with wires up to 100 feet long. Heavy-duty models with special motors and heavy cables operate reliably up to 200 feet from the electric outlet.

PAINTS

Oil-Based Paint. An unopened can of oil-based paint has a life-span of up to 50 years. The paint will separate over time, but it can be remixed. The paint's drying time may lengthen, even double, however, because of the evaporation of the paint's drying agent. All paint should be kept in a cool place, since excessive heat can cause the paint to fume and pop the can's lid with explosive force.

Once the can is opened, oil-based paint will react with air to form a skin over the paint. However, the paint can be kept for many months, even years, if certain precautions are taken; opened paint, for example, should be transferred to a close container. In other words, if you have a quart of paint left from a gallon can, transfer it to a quart container. A thin layer of linseed oil can also be spread over the surface of the paint to protect it from the air.

Water-Based Paint. Water-based paints have shorter life-spans than oil-based paints. At best they can be kept for 6 or 7 years unopened before they begin to deteriorate seriously. Water-based paints should be kept away from cold as well as heat because they can freeze, thus ruining the paint. Once opened, they also are susceptible to mildew or mold. If you store opened paint in a close, airtight container it will last for several months.

VARNISH

An unopened container of varnish can be kept an indeterminate period of time. Bottles of perfectly usable varnish, for example, have been recovered from sunken sailing ships after centuries at rest on the ocean bottom. Like paint, varnish should be kept in a cool place and, once opened, should be stored in a closed, airtight container.

SHELLAC

Shellac is, in fact, a varnish made from the resin of the lac insect. In has the same lengthy life-span as other varnishes.

CHAMOIS

Ideal for drying and polishing automobiles, these soft, pliant leather cloths originally were fashioned from the skin of the chamois, a small European antelope. These goatlike creatures are extremely rare today, and almost all "chamois" now sold are actually oil-tanned sheepskin. Both true chamois and the sheepskin imitations will withstand as much as 10 years of regular use if they are properly maintained. They should not be left in the sun or near direct heat; their oils will dry, and the leather will crack. They should be rinsed thoroughly in warm water after every use.

CHARCOAL LIGHTER FLUID

Charcoal lighter fluid will last indefinitely if unopened. It should be stored in a cool place and can be left out of doors since it will not freeze. But keep it in a sealed container—the lighter fluid will evaporate if the container is left open.

INSECTICIDES

Most insecticide manufacturers recommend that their products be stored no longer than 5 years, but theoretically they could be kept for centuries. They will, in fact, become stronger over time, as their solvents slowly evaporate. The majority of insecticides are oil-based; they therefore are highly flammable and should be kept in a cool place. Aerosols especially should be kept away from heat, for they can develop high pressures and explode. Dry or powdered insecticides are affected by moisture and should be kept in plastic bags or airtight containers. That convenient space under the sink is a poor place to store insecticides—or just about anything else, for that matter—for it is both warm and humid.

There are a few water-based insecticides that can be ruined by freezing. As for any poison, read the label carefully and store out of the reach of children.

Sports, Games, and Hobby Equipment

Sporting equipment is designed to be used and abused. The fundamental axiom of the sporting manufacturers seems to be "the tougher the sport, the sturdier the equipment." Still, it must be recognized that all equipment has to be geared toward the wear and tear placed upon it by the "average" player—the one who neither polishes his tennis racket to a high gloss after each match nor smashes it into a chain-link fence every time a point is lost.

The life-span of a piece of sporting equipment depends upon several factors: the quality of the materials used in construction, the extent of use, the degree of abuse, and the fussiness of the owner. Professional athletes, for example, use equipment only until the moment when it loses its capacity to perform perfectly; because their livelihoods depend upon striving for an ideal, they are willing to sacrifice durability for performance. The money the pro spends on a new tennis racket, for example, may result in power gained on a crucial serve. But the amateur has somewhat different standards. Playing with a slightly bloated football four times a year may be an inconvenience, but it is certainly no catastrophe. And a flubbed backhand in tennis or a missed jump shot in basketball usually has far less to do with the quality of the equipment than with that of the player.

Most sporting equipment, then, has two possible life-spans: the short but glorious one of professional sports, and the longer, use-it-till-it-falls-apart one of amateurs.

FOOTBALL

Footballs. Although it is called a pigskin, the football the professionals use is actually made of cowhide. The life-span of a professional ball depends upon whether it is used in practice or in a game. Practice balls last from 2 to 3 days—a playing life of perhaps 5 hours. Game balls in the National Football League have markedly shorter life-spans. Because the home team is required to provide 24 new balls for each game, and because 8 to 12 of these balls are actually used—and then discarded, given away, or sent to the practice field—a ball could be said to last about 6 minutes of playing time.

Footballs used by amateurs are available in either leather or rubber and can last for years, depending upon their use and the whims of the owners. Deflating them slightly before storing them adds to their durability, since full internal pressure eventually stretches their seams. Of course, this practice may also put stress on the owner and lead to a premature coronary, since reinflating the ball means that you have to dig through every closet and storage space in the house to find the tiny pin you need to attach the ball to the pump.

Helmets. The life-span of a professional football helmet is 3 to 4 years. Helmets could be used indefinitely if they didn't tend to crack at the point where the face mask attaches to the plastic shell.

Jerseys. The pros use mesh jerseys (tear-aways are about to be declared illegal), which last through one season plus the following exhibition season. In college football, the tear-away jersey may last only half a game. Earl Campbell of the University of Texas, the 1977 Heisman Trophy winner, may well hold the record for wearing jerseys with the shortest life-spans. In the 1977–1978 season, Campbell had 90 jerseys torn off his back—an average of one every 8 minutes. Tear-away jerseys cost $8 apiece.

Pads. Shoulder pads take a real beating, but nonetheless average two to three professional seasons. Hip, knee, and thigh pads last indefinitely.

Cleats. Professional cleats last 1 year. Amateurs—usually kids—wear them until they are outgrown.

BASKETBALL

Basketballs. Professional teams purchase approximately 30 balls per year. Loss, theft, and wear and tear from practice and games eliminate approximately three to four a month. The life-span of an amateur's basketballs depends upon the quality of their construction and whether they are bounced upon hardwood or cement. Most of them seem to be used until their skins are so thin that they develop blisters (small, asymmetrical outpockets of air); it then becomes impossible to dribble them.

Nets. Practice nets are changed every 2 to 3 weeks. Game nets in Madison Square Garden are changed twice a year.

Sneakers. The pros change their sneakers about once every 2 weeks. A schoolboy may use the same pair for a full year or more for everything from basketball to wading in the surf.

BASEBALL

Baseballs. Anyone who has played sandlot baseball with a ball that is more electrical tape than horsehide knows that the life-span of a baseball depends on its users. In major league baseball a ball could theoretically be used for an entire game, but this is never done. The replacement of balls—because they are dirty or nicked—is left to the discretion of the umpire, and with the added loss attributable to foul balls and home runs, it is not unusual for more than 100 balls to be used in a single game.

Baseballs have leather covers and thus should not be stored in dry, hot places where they will flake and crack. They should be used only on grass surfaces; surfaces such as asphalt or cement will scar them mercilessly.

Gloves. A top-line baseball glove could last a player's entire professional life if properly maintained. In practice, in the major leagues, a fielder's glove has an average life-span of two seasons. The catcher's mitt seldom lasts more than a single season.

Baseball gloves, like all leather products, should not be stored in a hot, dry environment. They should be kept clean and regularly treated with a leather lubricant, especially *inside* the glove, where salt left from perspiration will cause the glove to crack. New gloves must be broken in before they can be used at their best. Some players suggest placing a softball in the pocket of a new glove, binding the glove securely around the ball with twine, and then soaking the glove in water for up to a week to provide a permanent pocket. Gloves should always be carefully lubricated or conditioned after this treatment.

Bats. The life-span of a baseball bat can range from a single pitch to seasons of regular play. In general, bats do not last as long as they used to. This is because bats made today are not cured as thoroughly as they once were and because some of the more recently developed pitches—the slider, for example—often result in cracked bats. In the major leagues it is rare for a bat to last a full season.

Bats can be maintained in a number of ways. Some players "bone" their bats; that is, they rub the bat with a metal pipe, supposedly to close the bat's pores and increase its life-span.

Others store their bats in linseed oil during the off season to cause the wood to knit.

Uniforms. Major league baseball teams provide their players with two to four uniforms each season, depending on the wealth of the club. If a player uses up his supply he must then purchase his own. Curiously, the shirts wear out more quickly than the pants.

Shoes. In the major leagues shoes last an average of 2 months. What happens is that the cleats wear down. To maintain baseball shoes, carry them to the playing field. Do not walk with cleats on hard surfaces (which wear down the cleats like sandpaper), and do not bang the cleats with a bat to clean them (it is hard on the bat and dangerous for the foot).

Bases. Traditionally, bases were canvas bags filled with such things as straw, feathers, or shredded plastic. Today, although they are still referred to as "bags" or "sacks," most bases used in professional baseball are made of single pieces of hard rubber. For a game at Yankee Stadium three sets of these rubber bases—called Hollywood bases—are used: One set for practice, one set for the first four and a half innings, a third set for the remainder of the game. Although they are replaced three times a game, these hard-rubber bases are reused and can have life-spans of up to 3 years.

Home Plate. Made of hard rubber, home plate at Yankee Stadium is changed twice a season.

Pitcher's Mound and Rubber. The pitcher's mound in a major league baseball stadium is 100 percent clay (as is the area where the batter and catcher stand). The mounds are patched daily and will last an entire season. The hard-rubber bar embedded in the pitcher's mound where the pitcher places his pivot foot is replaced two to three times a year. The rubber has four sides and is rotated so that each face of the rubber can be used before replacement.

HOCKEY

Pucks. These hard-rubber disks are virtually indestructible, although they are occasionally chipped by a sharp skate blade or scarred by the steel-mesh nets, according to Jimmy Young, equipment manager of the New York Rangers of the National Hockey League. Even so, their life-spans in the NHL are incredibly short. The Rangers, for example, may use 40 pucks a game, all of them slapped over the boards and into the stands (even though hockey pucks are frozen before they are put into play in the NHL so that they will bounce as little as possible). In a season of practice and play, the team may go through 300 to 400 dozen.

Hockey Sticks. Most sticks are made of hardwood, usually ash, but they are easily broken. While their life-spans vary considerably, the average in the NHL is about two games per stick. In amateur games, where slapshots and stick checks are less fierce, a stick may last a season.

Skates. Pros use two pairs per season. With a 26-week season encompassing 6 hours of hockey a week (5 hours in practice plus 20 minutes on ice per game for three periods), a pair lasts about 78 hours of skating time. Of course, for the amateur, skates are barely broken in by then.

Goalies' Masks. Masks rarely break and can last a career.

Nets. With constant repair, a net can last a season.

TENNIS

Tennis Rackets. The life-span of a racket depends on how well—or roughly—it is used, how it is cared for, and how tightly it is strung. Professionals wear out four to five rackets in a summer—an average of about two tournaments per racket. But they string their rackets far more tightly than the average player, and the natural abuse a racket receives after missed shots, which tends only to scar the racket of an amateur, can destroy a pro's racket. Billie Jean King, for example, has been known to slam her racket on the court upon occasion; at times, it has simply collapsed from the strain.

Wood and metal rackets seem to last about the same length of time, although wood rackets require more care. If they are left untended, wood rackets can dry out and warp, while metal rackets begin to buckle and bend. Graphite rackets, the latest sporting toy on the courts, are too new to allow an accurate assessment of their life-spans.

Strings. The exact life expectancy of the strings on a racket depends upon the type of player (hard hitter, drop-shot artist), the surface played on (asphalt gives the ball a stronger, livelier bounce, and therefore places more strain on the strings), and how the racket is cared for. According to *Tennis Magazine,* "for the player who's on the court once a week, [strings] should be good for a season of at least six months." But for those of us who are a little less concerned, strings last the life-span of the racket.

Tennis Balls. Quality tennis balls are packaged in threes, in cans, under negative pressure or vacuum. Stored at room temperature, unopened, a can of tennis balls has a maximum life-span of 6 months. Once opened, the balls will deteriorate even more quickly.

The reason an unplayed tennis ball has such a short life-span is that it is produced with an internal pressure that is higher than normal atmospheric pressure. Over time, the internal pressure decreases and the ball loses some of its bounce. (Similarly, if a tennis ball is exposed to high temperatures, its internal pressure will increase and the ball will become too lively.)

In tournament tennis six new balls are usually introduced during the warm-up and are used for the first seven games.

After the seventh game the balls are discarded, usually sold to spectators, and new balls introduced. Balls are then changed every nine games. On more abrasive surfaces—coated cement, for example—balls are changed after the fifth game and every seven games after that. At tournaments held at high altitudes—Las Vegas, for example—the balls are usually opened 24 hours in advance so they will have a chance to adjust to local pressure conditions.

Shoes. Although the pros change their sneakers about once every 10 weeks, most are affiliated with the manufacturers and can get them free. For those of us who have to buy sneakers, a pair can last for years, depending upon what it is used for, how it is worn, how well it was designed and constructed in the first place, as well as how fussy its wearer is.

Court Surfaces. Clay courts need a lot of care, but can then last indefinitely. Grass courts, such as those at Wimbledon, survive a week under heavy play; then they have to be taken out of service and reseeded. Asphalt can last for 5 to 7 years before it begins to heave and buckle, but poor weather conditions—harsh winters and the like—can shorten its life-span considerably. Composition courts last well over 10 years, although they must be repainted every 4 or 5 years. Certain contractors will guarantee composition courts only if metal rackets are fitted with racket guards, since a fierce competitor trying to put away a drop shot can do more damage to the court's surface than to his or her opponent.

Nets. Nets last an average of 4 years. Their life-spans are improved if the tension on them is eased when they are not being used.

Lines. Painted lines, whether they are brushed onto asphalt or composition surfaces, last about 4 years. The cloth lines tacked onto clay courts average about 3 years of moderate play if they are cared for.

SQUASH

Squash Rackets. Practically every racket sold carries a variation on the following caveat: "Due to the nature of the game, squash rackets cannot be guaranteed against breakage." Squash is, indeed, a game that is rough on equipment. With its tiny space, unforgiving walls and floor, rocketlike ball, and lunging combatants, it practically demands that things break and be replaced. The average player who avoids most collisions with walls, ball, and opponent can expect a racket to last about 6 months; most tournament players and those who are blind to everything but the point have to change rackets about once every 45 days.

Squash Balls. The new American 70+ ball, by now the standard squash ball, will survive 1 to 2 weeks of heavy play and 1 month of moderate to easy play before it cracks or loses its resilience.

Courts. Traditional squash courts, constructed of maple slats, seem to last a lifetime if they are maintained; some are in excellent condition after 50 years of play. The newer courts, composed of either a maple-plywood amalgam or Masonite, will not wear as well: The wood courts will last perhaps 20 years, but the Masonite courts begin to deteriorate after only 2 years in some cases.

Court Shoes. The life-span of the shoe depends upon whether it is made to be durable or to be light. Squash shoes with gum-rubber soles will usually withstand 3 months of heavy play and more than a year of light play.

GOLF

Golf Clubs. "The life-span of a set of clubs depends upon the mind of its owner," notes one experienced observer of the golfing scene. The reason for this is clear: A golf club simply does not wear out if it is properly maintained and if its owner can avoid wrapping it around the nearest tree. Woods, which would seem to be susceptible to wear, can be refinished ad infinitum. And irons are virtually indestructible; even an iron that has to endure the ravages of an energetic 6-year-old beating it on rocks can be machined down, with a small weight added to restore its balance.

The potential life-span of a good set of clubs is reflected in the fact that some pros use sets they have had for decades, just because the feel is right. On the other hand, hordes of gadget-oriented duffers are certain to turn their sets in for newer models every couple of years, on the theory that technology, not practice or skill, will improve their games.

Golf Balls. Many professionals consider the life-span of a golf ball in competition to be one hole, whereas an amateur might use a ball over and over—at least until it takes a dive into the nearest water hazard. Even though the professional, with his accuracy and smooth swing, damages or loses balls less frequently than the amateur, he changes balls because a golf ball changes shape after impact and takes a while to become spherical again.

RUNNING AND JOGGING

Shoes. The life-spans of running shoes are measured not in time but in mileage. Still, the variation is incredible. One runner took on the trails of the eastern Sierra and had his shoes disintegrate after just 7 miles; another, finding a similar pair so comfortable that he wore them through two of their three soles, claims he logged 1,500 miles—on asphalt.

The life-span of a pair of running shoes depends on a multitude of factors: the running surface (hard concrete or soft grass), the runner's weight, the smoothness and length of stride, and the kind of stride—whether it is flat-footed, distributing impact and friction, or whether the heel or toe hits first.

It also depends on whether the shoes were constructed for durability or shock absorption: Shoes that absorb shock have soles made primarily of rubber; shoes made to last have composition soles of rubber and varying amounts of polyurethane. In any case, the average shoe worn by the average runner on a surface averaging somewhere between hard and soft will last from 350 to 500 miles.

Clothing. Durability has little to do with running togs; their life-spans bow to the vagaries of fashion.

SPORTS ARENAS

The 12,500-seat enclosed arena in the Hartford, Connecticut, Civic Center lasted only 3 years; then its 1,400-ton, 2.5-acre roof collapsed under the weight of a moderate snowfall in January 1978. In contrast, the 45,000-seat Colosseum in Rome, built in the first century A.D. and now a little over 1,900 years old, is still there for the curious to see.

SYNTHETIC TURF

Synthetic turf such as AstroTurf has a life-span of from 7 to 10 years outdoors or up to 20 years indoors. The playing field of the Houston Astrodome, for example, has had the same synthetic turf in use since 1965. Synthetic turf does not fade (the dye is contained inside the nylon fiber) and is impervious to water. Ultraviolet rays from the sun, however, can cause the fibers to change molecularly—the primary reason the turf lasts longer indoors.

GUNS AND AMMUNITION

Outside the army, firearms have indefinite life-spans. It is not uncommon for a hunter's rifle or shotgun, for example, to be passed on from generation to generation. A rifle used in the revolutionary war, if properly maintained, should still fire today. It would not be a terribly efficient weapon, but then it was not terribly efficient when it was first employed. The principal danger to a rifle or handgun is rust, so the weapon should always be kept lubricated.

Stored in a cool, dry place, ammunition can be kept for more than 50 years. The garage or basement is often too damp, especially for shotgun shells with paper cases that can absorb moisture. (Most shotgun-shell cases are now made from plastic, but paper shells can still be purchased.)

With the exception of .22 cartridges, most ammunition can be repacked, recapped, and then reused. Cartridges with metal cases can be refilled as many as 100 times.

DARTS

The sport of dart throwing dates back to a group of English archers who devised a game in which shortened arrows were thrown at the rounded or butt end of a wine cask. The life-spans of these makeshift darts were obviously short. Today, a set of darts may well outlast the player who uses them. Only the feathers, or flights, will wear out.

There are three parts to a modern dart—the point, the shaft or barrel, and the flight. Points are made of hardened steel and the only maintenance they require is sharpening. Surprisingly, dart points can be too sharp, in which case they will tend to bounce off rather than slide by the dart board's wire dividers. Barrels can be made of wood, plastic, or a variety of metals. Tungsten is extremely popular, for it allows a slim design (so that the darts can be thrown closer together) without sacrificing the weight needed for accuracy.

The most vulnerable part of a dart is the flight, usually feathers or plastic. The biggest danger to flights is what is called "Robin Hooding"—when a dart in flight strikes another already in the board. Both plastic and feather flights can be easily replaced, and some serious players replace them as often as every day. Feather flights, which are easily crushed, can often be rejuvenated by steaming. Darts with wooden barrels and/or feather flights should be stored in a warm, humid place for maximum life-span.

Dart Boards. The boards most often endorsed by dart organizations and favored by serious players are made from tightly packed bristles, usually hemp, mounted on a durable backing. The life-span of such a board is a function of use, but the boards are virtually indestructible. One bristle board, used nonstop 24 hours a day at a 30-day tournament in London's Rose and Crown Pub, is still in regular use at the office of the Accudart company. Bristle boards can dry out in warm, dry climates, but are otherwise damaged only by use. Cork boards are popular, but they are relatively expensive and have short lives. Boards also are made from compressed paper, but these will not hold up under heavy use. The earliest dartboards were made of hardwood, usually elm, and a few of these can still be obtained. Not only are they handsome, they also can outlive their owners. The only drawbacks are weight (up to 20 pounds) and the fact that they must be soaked in water daily.

Tournament dart boards are divided into 20 pie-shaped segments by a wire assembly called a spider. On some of the better boards the numbers that mark the various segments are removable, allowing the board to be rotated and thus prolonging its life.

PLAYING CARDS

The life-span of a deck of playing cards depends upon several factors: the playing surface, the body chemistry of the players (whether their hands are wet and clammy or dry), the way the cards are handled, and the fastidiousness of the individual who has the power to throw them out. Some decks can last a lifetime of more or less regular use—if the players do not object to greasy, dog-eared cards. Others, used in professional gambling casinos, last from 2 to 5 hours of play, after which they become worn and misshapen enough to slow the pace of the dealer.

TOYS

Few of man's products have life-spans as variable as his toys. Century-old lead soldiers, cast-iron fire engines, and mechanical banks are almost common. Dolls have survived from the early 1800s, even though made of fragile porcelain or bisque. And a growing number of antiques dealers thrive on the sale of toys and other artifacts based on Mickey Mouse, Flash Gordon, and other heroes of the 1920s, 1930s, and 1940s. Yet a toy rarely survives a year with all its vital parts unbroken and unlost. In fact, according to the only available estimate, nearly 1 toy in 10 sold in the United States is damaged or destroyed in less than a week of use.

The problem is not so much that manufacturers deliberately design their toys to be broken, opening the way for replacement purchases, as that they have no reason to build more durable playthings. Toys are not really made to be played with. They are made, like any other product, to be sold. Many are made to be sold on television. They must be dramatic and full of enough motion, color, flashing lights, and whirring buzzers to fill a 30-second commercial. Unfortunately, that does not guarantee that the toy will be fun to play with for longer than 30 seconds. It need not last for years, because quite often the child will have lost interest in minutes, a few days at most.

At the other end of the scale are electric trains, tricycles and bicycles, and a few favorite dolls and stuffed animals. Electric trains, trikes, and bikes are all made for hard use, at least if you compare them with plastic rockets and machine guns. Most tricycles survive until they are outgrown. Bicycles and trains are not only durable, but repairable as well. As long as replacement parts are available, they can be kept in use. Dolls and stuffed animals have just the opposite advantage: They are not sturdy, but the whole point of having a doll is to take care of it. Though many dolls and animals lose arms, legs, or stuffing, the favorites last as long as any other toy. For these playthings, a 10-year life-span is reasonably common. Most are outgrown, not destroyed.

PINBALL MACHINES

Although modern pinball machines are built to last longer, their novelty wears off after 2 years and they are replaced. Thereafter, many machines are rebuilt and sold to private owners, who can keep them operating for decades.

Pharmaceuticals, Prosthetic Devices, and Physical Aids

VITAMINS

In general, vitamins have life-spans of 3 years—vitamin E a little longer, vitamin C a little less. Natural vitamins have the same life-spans as synthetic vitamins except when the natural vitamin is carried in an organic medium—ground almonds, for example—and the medium spoils before the vitamin.

Capsules have slightly longer life-spans than tablets because they protect the vitamin better from the air. Vitamins in liquid form occasionally separate, but they can be remixed without loss of potency.

Vitamins should not be stored in the refrigerator. Many vitamins are water-soluble and will deteriorate in the humid environment of a refrigerator. Similarly, the bathroom is too moist. It's best to store vitamins in a dry place, at room temperature.

Vitamin manufacturers are not required by law to put expiration dates on their products, although many do. (There is currently a bill before Congress that will require vitamins to carry expiration dates.) As with cosmetics, the expiration date is occasionally marked on the box, but not on the actual bottle.

OVER-THE-COUNTER DRUGS

Not all over-the-counter drugs are required by law to carry expiration dates, but those that are must display the date on the actual container holding the drug, not merely on the carton or wrapper.

Frequently, two formulations manufactured by different companies will contain the exact same ingredients but will carry different expiration dates. What has happened is that the Food and Drug Administration stipulates a *maximum* expiration date for a particular product. The manufacturer, for obvious marketing reasons, may choose to display an earlier expiration date.

The life-spans of over-the-counter drugs range from 24 to 60 months. Aspirin, for example, is usually given a life-span of 48 months. However, it's extremely vulnerable to heat, light, and moisture and so should be stored in a dry, dark, cool place—as should all drugs. (The refrigerator actually is too cool; the bathroom, too warm and moist.) Any change in color or swelling of the tablets is a sign the aspirin is deteriorating.

Often the life-span of a drug is determined by the form in which it is packaged. One major manufacturer rates nasal drops at 36 months, while nasal spray is given a shelf life of 60 months. The same manufacturer produces an antifungal formulation with a life-span of 24 months in aerosol-powder form, 36 months as a cream, 60 months as a powder. Following the manufacturer's expiration date is the best idea, even though the dates often appear whimsical.

There is no sure way to tell whether an over-the-counter drug has deteriorated or spoiled, but any drug that changes color, clouds, or separates should be discarded unless the label specifies it can be kept in that condition.

PRESCRIPTION DRUGS

Pharmacists frequently do not fill in the expiration dates on prescription drugs because it is usually assumed that the customer will take the drug according to his physician's instructions—four times a day, or whatever—until the bottle is empty. On the chance that all the prescription may not be used immediately, you should ask the pharmacist to include the expiration date. However, do not begin to take any prescribed medication again without first consulting your doctor.

SLEEPING PILLS

Although sleeping pills have been in common medical use since the 1930s, it wasn't until 1970 that anyone bothered to find out just how long they remain effective. The person responsible for these measurements was Penn State psychiatrist Dr. Anthony Kales. Among the discoveries he and other researchers have made in recent years about prescription sleeping pills are that most give you an unnatural slumber that deprives you of your REM (rapid eye movement) sleep, when you do most of your dreaming; all induce withdrawal symptoms that usually include a backlash orgy of dreams or nightmares once you quit taking a pill; and, taken long enough, sleeping pills can *cause* insomnia, actually intensifying sleep disorders.

Because of variations in an individual's metabolism, Kales was unable to state exactly how long a pill worked over a period of a few days, but he was able to give general life-spans of effectiveness in terms of weeks. The accompanying table shows how long some of the most commonly used sleeping pills will work at putting you to sleep and keeping you asleep if you take them every night. Generic drug names are given first: specific brand names are given in parentheses.

Drug	*Time effective (weeks)*
Scopolamine (Sominex)	0*
Secobarbital (Seconal)	2 or less (usually less)
Pentobarbital (Nembutal)	2 or less (usually less)
Chloral hydrate (Noctec, Aquachloral)	2 or less (usually less)
Glutethimide (Doriden)	2 or less (usually less)
Methaqualone (Quaalude)	2 or less (usually less)
Ethchlorvynol (Placidyl)	2 or less (usually less)
Flurazepam (Dalmane)	More than 4†

*In tests, proved no more effective than a sugar pill.
†Dalmane was only pill that exceeded 2-week limit and was still working after 4 weeks of testing. Subsequent tests have indicated that its upper limit of effectiveness may be 5 or 6 weeks.

VACCINES

Vaccines and other immunizations against disease have been largely responsible for the remarkable increase in human life-span in the last century.

Although some vaccines provide lifetime protection against a particular disease, most lose their effectiveness over time. The accompanying table indicates how long you will remain immune to a disease once vaccinated:

Disease	Effective life-span of vaccination
Smallpox (cowpox virus)	5–10 years
Diphtheria (toxoid)	5–10 years
Diphtheria (antitoxin)	2–3 months
Typhoid (dead germs)	2–3 years
Whooping cough (dead germs)	2–5 years
Measles (attenuated virus)	Many years
Measles (immune blood serum, gamma globulin, or placental extract)	Few weeks
Mumps (attenuated virus)	Probably life
Tetanus (toxoid)	5–10 years
Tetanus (antitoxin)	Few weeks
Rabies (attenuated virus)	Unknown
Combined immunization (diphtheria and tetanus toxoids and whooping cough vaccine)	5–10 years
Poliomyelitis (dead or attenuated virus)	Unknown
Rubella (attenuated virus)	Unknown

PROSTHESES

When people think of prostheses, they usually envision Captain Ahab's wooden pegleg or the gun built into the artificial arm and hand of private investigator J. J. Armes. But prostheses include much more. The temporary plates that are screwed in place to protect and help heal nasty fractures, the artificial hips that are implanted in the elderly to replace deteriorating joints, the surgically installed rods used to straighten the curved spines in teenage girls are all prostheses as well.

There are really two categories of prosthetic life-spans: one for a prosthesis that is temporary and one for a prosthesis that is designed to last a patient's lifetime. A temporary prosthesis, like a fracture plate, is required to perform a specific function for a limited time. Sometimes such a prosthesis is installed for a few months, sometimes for years; sometimes it is simply left alone long after its useful life-span is over (the surgery required to remove it is often more painful and more dangerous than simply leaving it in place). The life-span of a temporary prosthesis, of course, depends, mainly on the job to be done, on the skill of the surgeon, and on the care taken by the patient.

The success of a prosthesis designed to last a lifetime—from the tiny, almost delicate replacements for the joints of the fingers to the more massive, weight-bearing ball-and-socket configurations for the hip—depends on the same factors that affect the temporary prosthesis. For the prosthesis to survive a human lifetime of wear, all factors must reach certain standards. A perfect prosthesis that is improperly placed during surgery, for example, can work its way free of its bony attachment, thereby requiring additional surgery; a prosthesis that is made for a patient weighing 150 pounds can easily fall apart if the individual gains an extra 100 pounds after surgery or if the bones in which it is placed begin to deteriorate; and a prosthesis that fails owing to faulty design or manufacture can foil the efforts of the most careful surgeon and patient.

It is important to remember that the lifetime a prosthesis must last is seldom the full life-span of the person in whom it is placed. Prostheses that must be surgically implanted are usually made for people whose joints are beginning to wear out—for the elderly or for those with bony malformations. Since these individuals often are in the latter parts of their lives, their implants may have to last only 5 or 10 years.

CONTACT LENSES

If they are carefully handled, contact lenses can last a long time. But invariably people who wear them lose, scratch, tear, or break at least one of their lenses long before its theoretical life-span has expired. As a result, the average wearer keeps a complete set of lenses for little more than a year before one of them has to be replaced.

Different kinds of lenses cause different problems. Hard lenses are more durable than soft lenses and last far longer. Gas-permeable lenses (a type of flexible hard lens that permits the flow of oxygen to the eye and carbon dioxide from it—a factor of comfort, and not of safety) last somewhere in between. Theoretically, under normal conditions, soft lenses can last about 2 to 3 years, while hard lenses endure at least twice that long. But while hard lenses are more durable, they are also smaller and more likely to drop out of the eye. Their actual life-spans, therefore, only minimally surpass those of soft lenses.

Research and development in contact lenses is progressing so rapidly that few people would care to keep their old lenses for their full life-spans, even if they could. Because prescriptions change and because better, safer, more comfortable, and more durable products are being invented almost yearly, many people decide to change voluntarily, long before they lose or destroy the lenses.

HEARING AIDS

A hearing aid is a fantastically miniaturized set of components. In a package not much larger than a thimble are packed microphone, amplifier, battery, and controls for volume and frequency.

Because so much is jammed into so little space, hearing aids are very delicate items. While the manufacturers give a theoretical life-span of 4 to 5 years for more durable aids (exclusive of the battery), it is assumed that people who treat them harshly or who wear them constantly will have to have them changed after only 3 years. On the other hand, people who take them home and leave them in dresser drawers until it is time for their next check-up can keep them indefinitely.

Hearing-aid batteries come in countless makes and sizes and are designed to fit the dozens of styles of hearing aids that exist. Their life-spans depend upon how often the hearing aid is used, the volume to which it is set, and the quality of the battery itself—a set of variables that makes it impossible to calculate any general longevity.

TOOTH FILLINGS

Any filling has to be replaced if the tooth continues to decay along the line where it meets the filling. Gold and porcelain fillings are fairly permanent, as are small silver fillings. Silver fillings that are used to cover large areas, however, frequently fracture when they are jolted or hit.

Fillings made of acrylic, composite, or cement, often used on front teeth for cosmetic reasons, are soluble in saliva and must eventually be replaced. There are several new plastic materials now used by dentists that can best be termed "semi-permanent"; that is, they are not vulnerable to saliva, but, like all plastics, are susceptible to mechanical wear and tear.

DENTURES

Dentures made of porcelain theoretically last forever; plastic dentures, which are subject to erosion, usually last as long as the wearer. All dentures have to be readjusted or replaced after a while, however, because the bone and tissue to which they are attached tend to degrade and change shape over the years, ruining the dentures' fit. This degradation, because it does not occur in people who retain their own teeth, is thought to be caused by the loss of the natural teeth; some scientists are now experimenting with ways of treating the problem electrically, transmitting complex electrical signals to the bone and tissue to try to "trick" them into behaving as if the natural teeth were still present. If the theory proves workable, the life-spans of dentures will be greatly extended.

PACEMAKERS

Pacemakers consist of two separate parts: a pulse generator, which contains the battery and circuitry, and an implantable lead—the wire and electrode attached directly to the heart.

The pulse generator, as a battery pack, has an easily determined life-span, but only for those models worn outside the body. In the past, once implanted in the body's hostile environment, the generator's longevity was considerably reduced; therefore, the theoretical longevity of a pulse generator differed considerably from its actual service life. But new technology—in the form of lithium-compound batteries that replaced the old mercury-zinc types—has allowed the pulse generator to be completely sealed off from its environment; today, the generator's longevity is beginning to approach the theoretical.

Pulse generators are made to last anywhere from 3 to 11 years or more. The difference in life-spans reflects a difference in cost: the cheaper the pulse generator, the shorter its life-span. Why would anyone want a cheaper pulse generator when his or her life is involved? Because many elderly patients who receive these implants are gravely ill, with several other serious underlying conditions. Their life-spans can often be estimated to within several months, and most doctors who perform the implant surgery feel it is a waste of money to put a pacemaker that will last 10 years into a body that will make it through only 2. So they put in a pulse generator that is calculated to last as long as the patient, adding a few years in case of error. Even if the patient outlives the pulse generator, the surgery needed to replace it is relatively easy—a simple incision, replacement of the original unit with a new one tucked into a layer of fat, and a little stitching. The heart need not be exposed a second time.

Implantable leads used to be delicate instruments susceptible to wear and tear in the stress-filled environment around the heart. Again, however, the technology has become more sophisticated; if the leads are placed correctly, they should last for the lifetime of the patient.

For Vanity's Sake

COSMETICS

Most cosmetics and toiletries have life-spans of 24 to 36 months. The manufacturers of these products are not required by law to display expiration dates, but many do. (Most products have a control or batch number on the actual container through which the expiration date can be traced.) Often, however, the expiration date appears on the outside box, but not on the bottle or jar itself, so it might be wise to date the actual container before throwing the box away.

Cosmetics are formulated to be stored at room temperature; they can be damaged by excessive heat, cold, or humidity. Contrary to popular wisdom, the refrigerator is too cold and can cause some liquid products to separate and lipstick to become brittle. The bathroom, especially the medicine chest, is too moist and hot.

Hair-coloring products are particularly delicate and, once opened, should be used within a few weeks. Their active ingredients are easily oxidized and their potencies quickly changed.

Such products as conditioners, rinses, shampoos, aerosol foam shaving creams, deodorants, and toothpastes are extremely stable and can be kept on the shelf, at room temperature, for up to 2 years. Products rich in oil—hand lotions and facial creams, for example—have shorter life-spans. Their oil eventually turns rancid, giving off an odor and sometimes causing the product to change color. Products in liquid form frequently change color or separate when exposed to excessive heat or cold. Unless otherwise specified on the label, they should be discarded.

Every time you dip your fingers into a bottle of foundation or a pot of eye shadow, germs are introduced into the preparation. Most commercial cosmetics therefore contain antibacterial agents, but mixing cosmetics together may reduce these agents' effectiveness. The accompanying table should provide a guide to how long some of the most common cosmetics may be stored.

Cosmetic	Life-span
Mascara	Best to keep only 3 months. After than, either the brush or the dried flakes, if contaminated, can cause infection in the eye.
Lipstick	Although returned by retail outlets to manufacturer after 3 months, lipsticks last for several years.
Powder	Indefinitely.
Nail polish	6 months.
Perfume	In an unopened bottle, not exposed to air, perfume never loses its smell. Opened, the fragrance fades over time; on skin, it lasts 3 to 4 hours.

ARTIFICIAL SUNTANS

Products that are designed to brown the skin without the help of the sun actually do work; the tanning effect is caused by the chemical dihydroxyacetone, which sets off a reaction in the skin that nobody yet understands.

Artificial suntans will last a week or more if they are treated carefully. If, however, you don't like your new color (skin has been known to become mottled, spotty, even yellow from the stuff), just scrub yourself three or four times with soap and water, and most of the color will be removed.

WIGS AND HAIRPIECES

Toupees, falls, wigs, and hairpieces, both the synthetic ones and those made of human hair, have life-spans of 2 to 4 years of daily wear if they are properly maintained. Most of the damage to a wig or hairpiece occurs to the base to which the hair is attached, rather than to the hair itself. Over time, the base, usually made of silk netting, will deteriorate and the hair will fall out.

A wig or hairpiece is best stored on a head-shaped canvas block when not in use and should be kept away from light, which will speed oxidation, especially of human hair. Oxidation causes the color of the hair to fade. A wig should be cleaned as seldom as possible, for cleaning shortens its life-span. Cleaning is usually done with a diluted dry-cleaning solution. A wig should also be brushed sparingly, as brushing tends to pull out hair. Therefore, a wig should not be worn to bed, since the matting that occurs during sleep will require extra brushing to repair.

FACE LIFTS

A successful face lift, performed upon average skin subjected to average stress and care both before and after the surgical procedure, lasts about 6 to 10 years. But almost nobody is average; the list of factors, both mental and physical, that determines how long a face lift might last for you is almost endless.

Mental factors center around the one crucial concept—body image. It is a concept that psychiatrists confront almost daily in its more esoteric implications. But it is also something that plastic surgeons deal with on a nuts-and-bolts level every time they touch their scalpels. Very few people want face lifts for medical reasons or, for that matter, to look like Liz Taylor. Most want nothing more than to become "normal" again, to make their mental images of themselves match the physical realities.

As a result, some face lifts last a lifetime. The people who have them wear them happily; but by the time they might be in line for renewals, their perspectives have changed. They are comfortable with themselves as they are; their life styles have changed and they no longer consider the physical aspects of their images as important; or perhaps they no longer think that the results are worth what they would have to spend. Whatever the reason, there are those who have a single successful experience and never go back for more.

Physical factors involve everything from the elasticity of the skin (an inherited factor) to the degree to which it has been used and abused over a lifetime. Skin that holds a face lift the longest has never been exposed to the sun or other elements (that peaches-and-cream complexion), is kept moist and clean, covers a healthy body, and is cared for after the surgery. Skin that reacts relatively poorly to the procedure has suffered the opposite treatment in every respect. Some additional factors that can damage the skin are smoking, alcohol, chronic illness, malnutrition, allergic reactions, and most important, the stresses of harsh cosmetic chemicals and the weather.

A successful face lift wipes about 10 years from one's appearance. The age of people having them used to average about 65, but has been decreasing over the past decade; repeat procedures are now quite common, with some people returning three and four times.

BREAST IMPLANTS

Implants are now composed of silicone bags filled with either silicone or a saline solution. The bags used to be attached to the chest wall; now they are surgically placed directly in the existing breast tissue and move with the breasts, keeping pace with the rest of the body as it grows old. The implants themselves are extremely durable and do not deteriorate with age or as a result of contact with the body chemistry. They are so strong, in fact, that, according to Dr. Barry Dolich, a New York plastic surgeon: "When people dig up our civilization millennia from now, they should find those bags intact."

Clothing

FUR COATS

Depending upon how it is maintained, an animal fur coat can last anywhere from a year to a lifetime. Rough treatment, which includes wearing shoulder bags (they rub the hair right off the coat) or sitting for long distances in cars (more friction—between the car seat and the coat), destroys even the most durable furs; careful cleaning, storage, and wear, on the other hand, can extend their life-spans significantly.

To best maintain a fur, avoid brushing or combing it; any friction will weaken the grip of the hair on the skin. Instead, simply shake it out a few times; animals do that all the time when they are wearing the fur, and it seems to work well enough for them. If your fur needs professional cleaning, be sure to take it to a specialist. Cleaners who concentrate on fur cleaning usually spin or agitate the fur in a tank partially filled with sawdust; the dirt clings to the dust and the fur comes out clean.

Furs should be placed in cold storage at 45° F when they are not going to be worn for a while. In any case, they should never be stored in plastic or other airtight materials, since they need to "breathe."

Some of the more durable furs include mink, sable, and raccoon. Fox, lynx, and the like are far more fragile, and deteriorate more quickly, no matter how well they are cared for.

SWEATERS

Any good sweater should last at least 5 years, given "normal use" and gentle washing and drying. With a very sedate owner, knitwear can last 20 years. Worn during heavy exercise, even the best sweater will succumb in 3 years or so. Wool sweaters and those of man-made fibers last about equally well; cashmere is a little less durable.

SHOES

Women's shoes run fairly predictable lives. Barring the occasional broken heel, most will survive the 1 to 2 years it takes fashion to decree their abandonment, then go unworn no matter what their condition.

Men's shoes, in contrast, offer mysteries rivaled in the clothing world only by the disappearance of socks from washing machines. Even in today's throw-away economy, some men can wear the same pair of shoes day in and day out for 15 years before the shoes give out. Others will, without visibly mistreating them, destroy an identical pair of shoes in less than a year. No believable explanation has ever been found for this phenomenon. For most men, shoes of reasonably good quality can be expected to last 3 years or so in regular use.

NYLON STOCKINGS AND PANTYHOSE

The manufacturers of women's hosiery claim that the abysmally short life-spans of nylon stockings and pantyhose are the fault of the consumer. Women, they argue, demand sheerness over durability, resulting in products that often don't last a single wearing. It does seem curious, however, that while technology can send a man to the moon and weave fibers smaller than a human hair that are still capable of lifting a luxury automobile, it cannot produce a pair of pantyhose that won't run.

Women's hosiery have indefinite shelf lives, so long as the package is not opened. They are extremely vulnerable to such things as air pollution, however, and literally dissolve if exposed to the heated exhaust fumes from an automobile. Runs in nylon stockings can be stopped by the application of clear nail polish, but the treated area will become stiff. Hosiery woven from lockstitched materials will not run but will deteriorate—holes instead of runs—nonetheless.

BRASSIERES

Howard Hughes once boasted that the brassiere he designed to accommodate Jane Russell's noteworthy endowments had been built according to the same engineering principles as a suspension bridge. He was entirely serious. The bridge's cables and towers are mirrored by the brassiere's straps and the wearer's shoulders.

No brassiere is going to last as long as even a short-lived bridge, but durability has improved with the development of synthetic fabrics. Broken straps and torn-out hooks, once a bane, seldom occur today. What damages modern foundation garments is exposure to body oils, perspiration, and repeated washing, all of which attack the elastic.

In theory, a brassiere could survive from 2 to 6 months of daily wear, depending on the quality of its construction. Long before that, however, accumulated perspiration would have rendered it uninhabitable. Given gentle hand washing and line drying, a brassiere can last as long as 2 years of intermittent use. Most are destroyed by machine washing and drying within 2 to 3 months. Those made of thin stretch fabrics endure about as long as others; padding tends to protect the elastic from oils and perspiration and extends the garment's life slightly.

ATHLETIC SUPPORTERS

Like other elastic garments, supporters are vulnerable to destruction by oils, perspiration, and machine washing. But since many of these garments are seldom laundered, the practical limitation on their service life is often the fortitude of the wearer. One way or the other, a 1-year-old supporter that has been used several times a week is nearing the end of its career.

FIG LEAVES

These traditional garments passed the height of their popularity before the development of modern polling and statistical methods, so hard data on the useful lives of fig leaves are unavailable. The few recent studies of their durability indicate that fig leaves begin to dry and become fragile in 3 days to 1 week, depending on the temperature and humidity of their environment. Their life-spans could theoretically be extended by applications of lanolin or petroleum jelly to retard loss of moisture and maintain pliability. In practice, however, fig leaves seldom survive even the party to which they are worn.

Government Issue

PATENTS

A patent granted to an inventor by the U.S. Patent and Trademark Office gives him the right to exclude others from making, using, or selling his invention in the United States for 17 years. That "invention" may be any new and useful process, machine, manufacture, composition of matter; any new and useful improvements in these categories; or even any distinct new variety of plant—any or all of which, with the Patent Office's approval, will be protected for the 17-year period.

TRADEMARKS

Trademarks—the names or symbols used by a manufacturer to distinguish his goods from those of others—last 20 years. Thereafter, they remain trademarks only if the goods they designate are still in use. Otherwise, they become a part of the public domain and may be used by someone else.

COPYRIGHTS

Practically any works of authorship that appear in a way the law describes as "fixed in any tangible medium of expression, now known or later developed, from which they [the works] can be perceived, reproduced, or otherwise communicated, either directly or with the aid of a machine or device" can be copyrighted. In plainer English, copyright protection extends to books, magazines, and other publications, as well as to music and lyrics, dramas, musicals, pantomimes, choreography, sound recordings, and anything that could be called audiovisual. According to the revised law, anything created or published after January 1, 1978, belongs to the author for the rest of his life plus 50 years after that unless he signs specific rights over to the publisher. Creations copyrighted before that date are protected for 28 years from the time of first publication, but when that 28-year period runs out, copyright can be renewed for 47 years more, giving a copyright life-span of a total of 75 years.

STAMPS

The practical life-span of a stamp—from the moment you drop it in the mailbox to the moment it arrives at its destination—is about 3 days. However, stamps have been known to have "transitory" life-spans as short as 18 hours and as long as 32 years—as one Brooklynite found out when he received a postcard that had been mailed to him from Manhattan over three decades earlier.

For collectors, a stamp's life-span is far longer. The first stamp issued—and therefore the stamp with the longest lifespan—was the British One Penny Black, which was distributed on May 6, 1840. In the first 6 months of the stamp's existence, 72 million One Penny Blacks were sold.

But being the oldest does not necessarily mean being the most expensive. That honor goes to the One Cent Black on Magenta of British Guiana (issued in 1856), a square inch of paper with dog-eared corners, a smudged postmark, and a badly rubbed surface that was sold in 1970 for $280,000.

PAPER MONEY

According to the Bureau of Engraving and Printing—which prints food stamps, customs documents, and invitations to White House functions, as well as currency—a $1 bill lasts about 18 months in circulation or for about 4,000 folds. Higher-denomination bills stay in circulation longer but, because they are made from the same quality of paper as singles, withstand the same number of folds.

United States paper money that has been worn or mutilated can be exchanged at face value as long as more than one-half of the original bill remains. If you have less than one-half the original bill, it can be exchanged only if you can convince the treasurer of the United States that the missing portion of the bill has been totally destroyed.

United States paper money is printed on high-quality paper with durable ink, and if you have any extra cash, you can store it quite effectively so long as you keep it away from moisture, flame, and rodents that feed on paper. You won't need much space—a bill is only 0.0043 inch thick and 1 million of them can be stacked in a cube that is less than 4 feet on a side. You may have to reinforce the floor, however, for 1 million bills weigh nearly 1 ton.

COINS

Coins are more durable than paper money, of course, and have considerably longer life-spans. When a coin is mutilated or sufficiently worn to be withdrawn from circulation, it is returned to one of several mints, where it is melted down and the metal reused.

The biggest danger to coins, however, is hoarding. One out of every seven pennies put into circulation, for example, ends up in piggy banks, old socks, and the bottoms of dresser drawers within a year. One out of every fourteen nickels shares a similar fate. (The attrition rate for higher denomination coins is zero.)

The penny faces another danger—cost. In 1963 it cost $.003 to produce a penny. By 1977 that figure had risen to $.007. Former Treasury Secretary William Simon estimates that by 1980 it might cost a penny to make a penny and that not long after that the government will be losing money on every one it mints.

U.S. ARMY LIFE-SPANS

The U.S. Army is perhaps the foremost defender of the tradition of maintenance in the world today. In true army fashion it has devised a four-tiered maintenance system to vigilantly prolong the life-spans of its equipment. At the bottom tier is Organizational Maintenance, performed by the individual soldier—cleaning, lubrication, minor adjustments and repairs, spit and polish. Next comes Direct Support Maintenance, performed by specialists at the division level, and then General Support Maintenance, carried out by still other specialists at the corps level. Finally, there is Depot Maintenance, entire installations where army equipment is repaired to what the army calls "like new" condition. The result of all this? The army gets maximum use from its goods.

The army rates its equipment in a complex fashion. Weapons are discussed in terms of MTBF (mean time between failure); ammunition in MRBF (mean rounds between failure). Most army equipment has two specified life-spans: a combat or mobilization life-span and a peacetime life-span.

Rifles. Army rifles are performance-rated for each particular model. The M-1, for example, which was the standard army-issue rifle for more than 20 years, was rated at 10,000 rounds. In general, army rifles are repaired or rebuilt so long as maintenance appears to be more economical than replacement.

Conventional Ammunition. The army claims that most types of conventional ammunition can be stored for 20 years or more. Ammunition is checked every 3 to 5 years to determine its serviceability.

Missiles. Missiles, according to the army, have shelf lives of from 10 to 15 years, depending on the type. After that time the missiles are inspected, and their life-spans may be extended. Antitank missiles, the army points out, have minimum shelf lives of 5 years, after which they are inspected periodically to ensure serviceability. Like other high-technology devices, missiles often grow obsolete before their theoretical life-expectancies are over.

Jeeps. For the current-model jeep, the army has established a life-span of 12 years or 72,000 miles, whichever occurs first.

Tanks. Because of their high cost, tanks are maintained as long as possible. A new generation of tanks seems to appear every 20 years or so, at which time the old tanks are replaced.

Parachutes. The life-span of a parachute is rated by the army at 15 years from the date of manufacture. This 15-year span is broken down into 3 years of storage and 12 years of operational use. The actual life-span of a parachute during this period is determined by regular inspection.

Uniforms. The mobilization life-span of an army uniform is 3 to 4 months; the peacetime life-span is rated at 15 months.

Boots. Leather combat boots have a mobilization life-span of 6 months, a peacetime life-span of 8 months. (The army walks during war and peace.)

Helmets. The mobilization life-span of a helmet is rated at 20 months, the peacetime life-span, at 32 months.

Canteens. The mobilization life-span of a canteen is 12 months, the peacetime life-span, 26 months.

Rations. C-rations (which replaced K-rations and are now themselves being phased out) have storage lives of 7 years. The army is now adopting freeze-dried rations, whose shelf lives it rates at 3 to 5 years.

Soap. Army soap has a life-span of 5 years in storage.

General Officers. What is the life-span of a general? The army says a general's career life, at least, is determined by the rank he eventually attains. The following life-spans indicate the career lives of various grades of general officer—in other words, the total time an officer spends as a general based on the highest rank he attains:

One-star general	3 years, 11 months
Two-star general	6 years
Three-star general	9 years, 6 months
Four-star general	11 years, 6 months

APPENDIX/LIFE-SPANS
OF THE FAMOUS

SOVEREIGNS AND OTHER ROYALTY

Ikhnaton	c. 1375–1358 B.C.	Isabella I	1451–1504
King Solomon	c. 973–933 B.C.	Henry VIII	1491–1547
Darius	c. 558–486 B.C.	Francis I	1494–1547
Xerxes	c. 519–465 B.C.	Anne Boleyn	c. 1507–1536
Alexander the Great	356–323 B.C.	Catherine de Medicis	1519–1589
Julius Caesar	100–44 B.C.	Ivan the Terrible	1530–1584
Herod	c. 73–4 B.C.	Elizabeth I	1533–1603
Cleopatra	69–30 B.C.	Mary, Queen of Scots	1542–1587
Augustus	63 B.C.–A.D. 14	Charles I	1600–1649
Caligula	A.D. 12–41 A.D.	Louis XIV	1638–1715
Agrippina the Younger	c. 15–59	Frederick the Great	1712–1786
Nero	37–68	Maria Theresa	1717–1780
Hadrian	76–138	Bonnie Prince Charlie	1720–1788
Constantine I	c. 280–337	Marquise de Pompadour	1721–1764
Justinian I	483–565	Catherine the Great	1729–1796
Charlemagne	742–814	George III	1738–1820
Eleanor of Aquitaine	c. 1122–1204	Ali Pasha	1741–1822
Richard the Lion-Hearted	1157–1199	Louis XVI	1754–1793
Genghis Khan	c. 1162–1227		
Kublai Khan	1216–1294		
Ivan the Great	1440–1505		

Marie Antoinette	1755–1793	Franz Josef	1830–1916
Joséphine de Beauharnais	1763–1814	Crazy Horse (Sioux Chief)	c. 1849–1877
Napoleon Bonaparte	1769–1821	Franz Ferdinand	1863–1914
Queen Victoria	1819–1901	Nicholas II	1868–1918
Prince Albert of Saxe-Coburg-Gotha	1819–1861	Haile Selassie	1892–1975
		Duke of Windsor	1894–1972
Compte de Chambord	1820–1883	King Faisal Ibn Abdul-Aziz al Saud	c. 1906–1975

MILITARY LEADERS

Pyrrhus	318–272 B.C.	William T. Sherman	1820–1891
Hannibal	247–182 B.C.	U. S. Grant	1822–1885
Attila the Hun	c./A.D. 406–453	Thomas ("Stonewall") Jackson	1824–1863
Genghis Khan	1162–1227		
Tamerlane	c. 1336–1405	George Custer	1839–1876
Sir Francis Drake	c. 1540–1596	Ferdinand Foch	1851–1929
Lord Cornwallis	1738–1805	Douglas MacArthur	1880–1964
Mikhail Kutuzov	1745–1813	George Patton	1885–1945
Horatio Nelson	1758–1805	Bernard Montgomery	1887–1976
Duke of Wellington	1769–1852	T. E. Lawrence (of Arabia)	1888–1935
Karl von Clausewitz	1780–1831	Erwin Rommel	1891–1944
Count Helmuth von Moltke	1800–1891		
Robert E. Lee	1807–1870		

STATESMEN AND HEADS
OF STATE

Pericles	c. 495–429 B.C.
Demosthenes	c. 385–322 B.C.
Cato the Elder	234–149 B.C.
Cicero (Marcus Tullius)	106–43 B.C.
Marcus Aurelius	121–180 A.D.
Duc de Richelieu	1585–1642
Oliver Cromwell	1599–1658
Jules Mazarin	1602–1661
Sir Robert Walpole	1676–1745
Grigori Potëmkin	1739–1791
Charles M. de Talleyrand	1754–1838
Klemens W.N.L. von Metternich	1773–1859
Daniel Webster	1782–1852
Benjamin Disraeli	1804–1881
Jefferson Davis	1808–1889
Hamilton Fish	1808–1893
William Gladstone	1809–1898
Otto von Bismarck	1815–1898
Porfirio Díaz	1830–1915
Sitting Bull	c. 1834–1890
John Hay	1838–1905
Georges Clemenceau	1841–1929
Arthur James Balfour	1848–1930
Cecil Rhodes	1853–1902
George Nathaniel Curzon	1859–1925
David Lloyd George	1863–1945
Mohandas Gandhi	1869–1948
Vladimir I. Lenin	1870–1924
Bernard Baruch	1870–1965
Cordell Hull	1871–1955
Winston Churchill	1874–1965
Konrad Adenauer	1876–1967
Joseph Stalin	1879–1953
George Marshall	1880–1959
Kemal Atatürk	1881–1938
Alcide de Gasperi	1881–1954
Benito Mussolini	1883–1945
David Ben-Gurion	1886–1973
Chiang Kai-shek	1887–1975
Adolf Hitler	1889–1945
Jawaharlal Nehru	1889–1964
Antonio de Oliveira Salazar	1889–1970
Charles de Gaulle	1890–1970
Francisco Franco	1892–1975
Dean Acheson	1893–1971
Mao Tse-tung	1893–1976
Nikita Khrushchev	1894–1971
Trygve Lie	1896–1968
Anthony Eden	1897–1977
Lester Pearson	1897–1972
Chou En-lai	1898–1976
Golda Meir	1898–1978
Dag Hammarskjöld	1905–1961
U Thant	1909–1974
Georges Pompidou	1911–1974
Gamal Abdel Nasser	1918–1970

NATIONAL HEROES

El Cid (Rodrigo Díaz de Bivar)	c. 1040–1099	Karageorge	c. 1766–1817
Alexander Nevski	c. 1220–1263	Davy Crockett	1786–1836
Joan of Arc	1412–1431	Sam Houston	1793–1863
Benjamin Franklin	1706–1790	Giuseppe Mazzini	1806–1872
Paul Revere	1735–1818	Charles Parnell	1846–1891
John Paul Jones	1747–1792	Emilio Aguinaldo	c. 1869–1964
Nathan Hale	1755–1776	Charles Lindbergh	1902–1974

REVOLUTIONARIES

Guy Fawkes	1570–1606	Roger Casement	1864–1916
Jean Paul Marat	1743–1793	Sun Yat-sen	1866–1925
Maximilien de Robespierre	1758–1794	Emma Goldman	1869–1940
Georges Danton	1759–1794	Rosa Luxemburg	1870–1919
Charlotte Corday	1768–1793	Emiliano Zapata	1879–1919
Simón Bolívar	1783–1830	Leon Trotsky	1879–1940
Nat Turner	1800–1831	Aleksandr Kerensky	1881–1970
John Brown	1800–1859	Bartolomeo Vanzetti	1888–1927
Giuseppi Garibaldi	1807–1882	Nicola Sacco	1891–1927
Mikhail Bakunin	1814–1876	Ernesto ("Che") Guevara	1928–1967

MERCHANTS AND CAPTAINS OF INDUSTRY

Cosimo de' Medici	1389–1464
Jacques Coeur ("The Merchant Prince")	c. 1395–1456
Jacob Fugger	1459–1525
Meyer Rothschild	1743–1812
John Jacob Astor	1763–1848
Eleuthère Irénée DuPont	1771–1834
Friedrich Krupp	1787–1826
Commodore Cornelius Vanderbilt	1794–1877
Andrew Carnegie	1835–1919
John Pierpont Morgan	1837–1913
John D. Rockefeller	1839–1937
Jacob Schiff	1847–1920
Henry Clay Frick	1849–1919
Andrew Mellon	1855–1937
Henry Ford	1863–1947
André Citroën	1878–1935
Helena Rubinstein	c. 1882–1965
Gabrielle ("Coco") Chanel	1883–1971
Elizabeth Arden	1884–1966
H. L. Hunt	1889–1974
J. Paul Getty	1892–1976
James Warburg	1896–1969
Howard Hughes	1905–1976
Aristotle Onassis	c. 1906–1975

JURISTS AND LAW ENFORCEMENT OFFICERS

Hammurabi	c. 1792–1750 B.C.
Sir Edward Coke	1552–1634
John Marshall	1775–1835
Roger Taney	1777–1864
Sir Robert Peel	1788–1850
Allan Pinkerton	1819–1884
James Butler ("Wild Bill") Hickok	1837–1876
Oliver Wendell Holmes	1841–1935
Wyatt Earp	1848–1929
Louis Brandeis	1856–1941
Charles Evans Hughes	1862–1948
Learned Hand	1872–1961
Felix Frankfurter	1882–1965
Hugo Black	1886–1971
Earl Warren	1891–1974
J. Edgar Hoover	1895–1972

PHILOSOPHERS, THEOLOGIANS, AND OTHER SOCIAL, AND SPIRITUAL REFORMERS

Thales of Miletus	c. 640–546 B.C.
Zoroaster	628–551 B.C.
Siddhartha Gautama (Buddha)	c. 563–483 B.C.
Confucius	c. 551–479 B.C.
Socrates	469–399 B.C.
Plato	c. 427–347 B.C.
Diogenes	c. 412–323 B.C.
Aristotle	384–322 B.C.
Epicurus	341–270 B.C.
Jesus Christ	c. 4 B.C.–A.D. 29
Lucius Annaeus Seneca the Younger	c. 4 B.C.–A.D. 65
Saint Augustine	A.D. 354–430
Mohammed	570–632
Saint Bernard of Clairvaux	c. 1090–1153
Maimonides	1135–1204
Saint Francis of Assisi	1182–1226
Jan Hus	c. 1369–1415
Martin Luther	1483–1546
Saint Ignatius of Loyola	1491–1556
John Knox	c. 1505–1572
John Calvin	1509–1564
Cornelis Jansen	1585–1638
Thomas Hobbes	1588–1679
William Rogers	c. 1603–1683
John Locke	1632–1704
Gottfried Leibniz	1646–1716

Voltaire (François Marie Arouet)	1694–1778
John Wesley	1703–1791
David Hume	1711–1776
Jean Jacques Rousseau	1712–1778
Denis Diderot	1713–1784
Immanuel Kant	1724–1804
Mary Wollstone-craft	1759–1797
Thomas Malthus	1766–1834
Georg Hegel	1770–1831
Brigham Young	1801–1877
Cardinal John Henry Newman	1801–1890
Dorothy Dix	1802–1887
William Lloyd Garrison	1805–1879
John Stuart Mill	1806–1873
Sören Kierkegaard	1813–1855
Frederick Douglass	1817–1895
Amelia Bloomer	1818–1894
Herbert Spencer	1810–1903
Susan B. Anthony	1820–1906
Mary Baker Eddy	1821–1910
Ramakrishna	1836–1886
Friedrich Nietzsche	1844–1900
Carry Nation	1846–1911
Samuel Gompers	1850–1924
Eugene Debs	1855–1926

Emmeline Pankhurst	1858–1928
Edmund Husserl	1859–1938
Henri Bergson	1859–1941
John Dewey	1859–1952
Jane Addams	1860–1935
William E. B. DuBois	1868–1963
Mohandas Gandhi	1869–1948

Bertrand Russell	1872–1970
Margaret Sanger	1883–1966
Paul Tillich	1886–1965
Martin Heidegger	1889–1976
Aimee Semple McPherson	1890–1944
Martin Luther King, Jr.	1929–1968

EXPLORERS

Marco Polo	1254–1324
Christopher Columbus	1451–1506
Amerigo Vespucci	1454–1512
Ponce de Leon	c. 1460–1521
Vasco da Gama	c. 1469–1524
Vasco Núñez de Balboa	c. 1475–1519
Ferdinand Magellan	1480–1521
Hernando de Soto	c. 1500–1542
Sir Francis Drake	c. 1540–1596

Vitus J. Bering	1680–1741
Captain James Cook	1728–1779
Daniel Boone	1734–1820
Meriwether Lewis	1774–1809
David Livingstone	1813–1873
Robert Edwin Peary	1856–1920
Roald Amundsen	1872–1928
Richard Evelyn Byrd	1888–1957

SCIENTISTS, INVENTORS, AND MATHEMATICIANS

Anaximander	611–547 B.C.		Georg Ohm	1787–1854
Pythagoras	582–500 B.C.		Samuel Morse	1791–1872
Democritus	c. 460–370 B.C.		Charles Babbage	1792–1871
Archimedes	c. 287–212 B.C.		Charles Goodyear	1800–1860
Hipparchus	190–120 B.C.		Justus von Liebig	1803–1873
Ptolemy	A.D. 90–168		John Louis Rodolphe Agassiz	1807–1873
Leonardo Fibonacci	c. 1170–1240		Charles Darwin	1809–1882
Johann Gutenberg	1400–1468		Henry Bessemer	1813–1898
Nicolaus Copernicus	1473–1543		Samuel Colt	1814–1862
John Dee	1527–1608		George Boole	1815–1864
Tycho Brahe	1546–1601		Gregor Mendel	1822–1884
Francis Bacon	1561–1626		Alfred Nobel	1833–1896
Galileo Galilei	1564–1642		Dmitri Mendeleev	1834–1907
Johannes Kepler	1571–1630		Ernst Mach	1838–1916
René Descartes	1596–1650		Karl Benz	1844–1929
Blaise Pascal	1623–1662		Wilhelm Roentgen	1845–1923
Robert Boyle	1627–1691		Alexander Graham Bell	1847–1922
Christian Huygens	1629–1695		Thomas Alva Edison	1847–1931
Anton van Leeuwenhoek	1632–1723		Luther Burbank	1849–1926
Sir Isaac Newton	1642–1727		Albert Michelson	1852–1931
Edmund Halley	1656–1742		George Eastman	1854–1932
Daniel Bernoulli	1700–1782		Konstantin Tsiolkovsky	1857–1935
Carolus Linnaeus	1707–1778		Max Planck	1858–1947
James Watt	1736–1819		Pierre Curie	1859–1906
William Herschel	1738–1822		Alfred North Whitehead	1861–1947
Antoine Lavoisier	1743–1794		Charles Steinmetz	1865–1923
Eli Whitney	1765–1825		Marie Curie	1867–1934
John Dalton	1766–1844		Ernest Rutherford	1871–1937
Jean Fourier	1768–1830		Guglielmo Marconi	1874–1937
George Stephenson	1781–1848			
John James Audubon	c. 1785–1851			

Albert Einstein	1879–1955	John Logie Baird	1888–1946
Niels Bohr	1885–1962	Werner Heisenberg	1901–1976
Clarence Birdseye	1886–1956	J. Robert	
Erwin Schrödinger	1887–1961	Oppenheimer	1904–1967

PHYSICIANS, PSYCHOLOGISTS, AND MEDICAL LUMINARIES

Hippocrates of Cas	c. 460–377 B.C.	Robert Koch	1843–1910
Paracelsus	c./A.D. 1493–1541	Wilhelm Roentgen	1845–1923
Edward Jenner	1749–1823	Walter Reed	1851–1902
Oliver Wendell Holmes	1809–1894	Sigmund Freud	1856–1939
		Havelock Ellis	1859–1939
Ignaz Semmelweis	1818–1865	Alfred Adler	1870–1937
Florence Nightingale	1820–1910	Carl Jung	1875–1961
Clara Barton	1821–1912	Albert Schweitzer	1875–1965
Louis Pasteur	1822–1895	Alexander Fleming	1881–1955
Joseph Lister	1827–1912	Karen Horney	1885–1952
Richard von Krafft-Ebing	1840–1902	Alfred Kinsey	1894–1956
		Aleksandr Luria	1902–1977

NOVELISTS, PLAYWRIGHTS, AND POETS

Aesop	c. 620–560 B.C.	Victor Hugo	1802–1885
Aeschylus	525–456 B.C.	Ralph Waldo Emerson	1803–1882
Euripides	c. 480–406 B.C.	George Sand	1804–1876
Aristophanes	c. 448–380 B.C.	Edgar Allan Poe	1809–1849
Vergil	70–19 B.C.	Charles Dickens	1812–1870
Ovid	43 B.C.–A.D. 18	Henry David Thoreau	1817–1862
Lucius Annaeus Seneca	c. 4 B.C.–A.D. 65	Fédor Dostoevski	1821–1881
Plutarch	A.D. 46–120	Gustave Flaubert	1821–1880
Dante	1265–1321	Leo Tolstoy	1828–1910
Francesco Petrarch	1304–1374	Lewis Carroll	1832–1898
Geoffrey Chaucer	c. 1340–1400	Mark Twain	1835–1910
Baldassare Castiglione	1478–1529	Emile Zola	1840–1902
François Rabelais	c. 1494–1553	George Bernard Shaw	1856–1950
Miguel de Cervantes Saavedra	1547–1616	Joseph Conrad	1857–1924
William Shakespeare	1564–1616	William Butler Yeats	1865–1939
Izaak Walton	1593–1683	H. G. Wells	1866–1946
John Milton	1608–1674	Luigi Pirandello	1867–1936
Molière (Jean Baptiste Poquelin)	1622–1673	Marcel Proust	1871–1922
Jonathan Swift	1667–1745	Robert Frost	1874–1963
Alexander Pope	1688–1744	Thomas Mann	1875–1955
Samuel Johnson	1709–1784	Jack London	1876–1916
Johann Wolfgang von Goethe	1749–1832	Virginia Woolf	1882–1941
Johann Fichte	1762–1814	James Joyce	1882–1941
Lord Byron	1788–1824	Eugene O'Neill	1888–1953
Aleksander Pushkin	1799–1837	T. S. Eliot	1888–1964
		Agatha Christie	1891–1976
		F. Scott Fitzgerald	1896–1940
		William Faulkner	1897–1962
		Bertolt Brecht	1898–1956
		Ernest Hemingway	1898–1961
		Robert Lowell	1917–1977

ARTISTS, ARCHITECTS, AND PHOTOGRAPHERS

Praxiteles	c. 370–330 B.C.
Giotto	c./A.D. 1266–1337
Donatello	c. 1386–1466
Sandro Botticelli	c. 1444–1510
Hieronymus Bosch	c. 1450–1516
Leonardo da Vinci	1452–1519
Albrecht Dürer	1471–1528
Michelangelo Buonarroti	1475–1564
Titian	1477–1576
Pieter Brueghel	c. 1525–1569
El Greco	c. 1541–1614
Peter Paul Rubens	1557–1640
Diego Velázquez	1599–1660
Sir Anthony Van Dyck	1599–1641
Rembrandt van Rijn	1606–1669
Christopher Wren	1632–1723
William Hogarth	1697–1764
Thomas Gainsborough	1727–1788
Francisco José de Goya	1746–1828
Jacques David	1748–1825
Katsushika Hokusai	1760–1849
Ando Hiroshige	1797–1858
Eugène Delacroix	1798–1863

Mathew Brady	1823–1896
Edouard Manet	1832–1883
Edgar Degas	1834–1917
Claude Monet	1840–1926
Auguste Rodin	1840–1917
Mary Cassatt	1845–1926
Vincent van Gogh	1853–1890
Stanford White	c. 1853–1906
Louis Sullivan	1856–1924
Anna Mary ("Grandma") Moses	1860–1961
Alfred Stieglitz	1864–1946
Henri de Toulouse-Lautrec	1864–1901
Henri Matisse	1869–1954
Frank Lloyd Wright	1869–1959
Aubrey Beardsley	1872–1898
Paul Klee	1879–1940
Edward Steichen	1879–1973
Pablo Picasso	1881–1973
Walter Gropius	1883–1969
Edward Weston	1886–1958
Le Corbusier (Charles Jeanneret-Gris)	1887–1965
Man Ray	1890–1976
Alexander Calder	1898–1976

HISTORIANS, ECONOMISTS, AND POLITICAL SCIENTISTS

Herodotus	c. 484–425 B.C.	Edmund Burke	1729–1797
Thucydides	c. 460–400 B.C.	Edward Gibbon	1737–1794
Livy	59 B.C.–A.D. 17	Thomas Paine	1737–1809
Pliny	A.D. 23–79	David Ricardo	1772–1823
Tacitus	55–120	Jules Michelet	1798–1874
Niccolò Machiavelli	1469–1527	George Bancroft	1800–1891
Sir Thomas More	1478–1535	Alexis de Tocqueville	1805–1859
Michel de Montaigne	1533–1592	Karl Marx	1818–1883
Thomas Hobbes	1558–1679	Heinrich Treitschke	1834–1896
Giovanni Vico	1668–1744	Thorstein Veblen	1857–1929
Adam Smith	1723–1790	John Maynard Keynes	1883–1946

DANCERS, CHOREOGRAPHERS, AND BALLET MASTERS

Marie Sallé	1707–1756	Sergei Diaghilev	1872–1929
Jean Georges Noverre	1727–1810	Ruth St. Denis	1877–1968
Carlo Blasis	1795–1878	Isadora Duncan	1878–1927
Maria Taglioni	1804–1884	Agrippina Vaganova	1879–1951
Auguste Bournonville	1805–1879	Rudolf von Laban	1879–1958
Fanny Elssler	1810–1884	Michel Fokine	1880–1942
Jules Joseph Perrot	1810–1892	Anna Pavlova	c. 1880–1931
Marius Petipa	1822–1910	Vaslav Fomich Nijinsky	1890–1950
Lev Ivanov	1834–1901	Doris Humphrey	1895–1958

ACTORS, ENTERTAINERS, AND FILM MAKERS

Fanny Brice	1891–1951
Richard Burbage	c. 1567–1619
Nell Gwyn	1650–1687
Edwin Forrest	1806–1872
Edwin Booth	1833–1893
Sarah Bernhardt	1844–1923
Max Reinhardt	1873–1943
Harry Houdini	1874–1926
David W. Griffith	1875–1948
W. C. Fields	1879–1946
Cecil B. DeMille	1881–1959
Douglas Fairbanks	1883–1939
Erich von Stroheim	1885–1957
Al Jolson	1886–1950
Jean Cocteau	1889–1963
Charlie Chaplin	1889–1977
Fritz Lang	1890–1976
Edward G. Robinson	1893–1973
Alfred Lunt	1893–1977
Lillian Leitzel	1894–1931
Jack Benny	1894–1974
Rudolph Valentino	1895–1926
Buster Keaton	1895–1966
John Ford	1895–1973
Busby Berkeley	1895–1976
Groucho Marx	1895–1977
Sergei Eisenstein	1898–1948
Paul Robeson	1898–1976
Humphrey Bogart	1899–1957
Spencer Tracy	1900–1967
Clark Gable	1901–1960
Walt Disney	1901–1966
Joan Crawford	1903–1977
Jean Harlow	1911–1937
Vivien Leigh	1913–1967
Gypsy Rose Lee	1914–1970
Lenny Bruce	1926–1966
Marilyn Monroe	1926–1962
James Dean	1931–1955
Freddie Prinze	1954–1977

SINGERS

Hans Sachs ("Die Meister-singer")	1494–1576
Carlo Broschi Farinelli	1705–1782
Jenny Lind	1820–1887
Adelina Patti	1843–1919
Feodor Chaliapin	1873–1938
Enrico Caruso	1873–1921
Luisa Tetrazzini	1874–1940
Geraldine Farrar	1882–1967
Sophie Tucker	1884–1966
Lauritz Melchior	1890–1973
Beniamino Gigli	1890–1957
Fanny Brice	1891–1951
Ezio Pinza	1893–1957

Bessie Smith	c. 1894–1937
Kirsten Flagstad	1895–1962
Rosa Ponselle	1897–1977
Jimmy Rodgers	1897–1933
Bing Crosby	1903–1977
Lily Pons	1904–1976
Woody Guthrie	1912–1967
Edith Piaf	1915–1963
Billie Holiday	1915–1959
Judy Garland	1922–1969
Maria Callas	1923–1977
Hank Williams	1923–1953
Jacques Brel	1929–1978
Elvis Presley	1935–1977
Janis Joplin	1943–1970
Jimi Hendrix	1943–1970

COMPOSERS, LYRICISTS, CONDUCTORS, AND MUSICIANS

Alessandro Scarlatti	1659–1725
Jean Philippe Rameau	1683–1764
Johann Sebastian Bach	1685–1750
George Frederick Handel	1685–1759
Franz Joseph Haydn	1732–1809
Lorenzo da Ponte	1749–1838
Wolfgang Amadeus Mozart	1756–1791
Ludwig van Beethoven	1770–1827
Francis Scott Key	1779–1843
Franz Schubert	1797–1828
Gaetano Donizetti	1797–1848
Hector Berlioz	1803–1869
Felix Mendelssohn	1809–1847
Fréderic Chopin	1810–1849
Richard Wagner	1813–1883
Giuseppe Verdi	1813–1901
Stephen Foster	1826–1864
Johannes Brahms	1833–1897
Sir William Gilbert	1836–1911
Pëtr Tchaikovsky	1840–1893
Sir Arthur Sullivan	1842–1900
John Philip Sousa	1854–1932
Giacomo Puccini	1858–1924
Victor Herbert	1859–1924
Gustav Mahler	1860–1911
Richard Strauss	1864–1949
Arturo Toscanini	1867–1957
Scott Joplin	1868–1917
Charles Ives	1874–1954
Pablo Casals	1876–1973
Bunk Johnson	1879–1930
Artur Schnabel	1882–1951
Igor Stravinsky	1882–1971
Leopold Stokowski	1882–1977
Otto Klemperer	1885–1973
Jelly Roll Morton	1885–1941
Cole Porter	1893–1964
Oscar Hammerstein II	1895–1960
George Gershwin	1898–1937
Edward Kennedy ("Duke") Ellington	1899–1974
Kurt Weill	1900–1950
Louis Armstrong	1900–1971
Bix Beiderbecke	1903–1931
Thomas ("Fats") Waller	1904–1943
Glenn Miller	1904–1944
Dorothy Fields	1905–1974
Coleman Hawkins	1907–1969
Woody Guthrie	1912–1967
Charlie ("Bird") Parker	1920–1955
Erroll Garner	1921–1955
Julian ("Cannonball") Adderley	1928–1975

ATHLETES AND SPORTSMEN

Abner Doubleday	1819–1893
John L. Sullivan	1858–1918
Walter Camp	1859–1925
Annie Oakley	1860–1926
Joe Darby	1861–1937
James Naismith	1861–1939
Connie Mack	1862–1956
Amos Alonzo Stagg	1862–1965
Bob Fitzsimmons	1862–1917
James ("Gentleman Jim") Corbett	1866–1933
Pudge Heffelburger	1867–1954
Cy Young	1867–1955
Tex Rickard	1870–1929
Harry Vardon	1870–1937
Reginald F. Doherty	1872–1910
Wee Willie Keeler	1872–1923
John McGraw	1873–1934
Ray Ewry	1873–1937
Honus Wagner	1874–1955
Sunny Jim Fitzsimmons	1874–1966
Hugh L. Doherty	1875–1919
Jim Jeffries	1875–1953
Frank Chance	1877–1924
Jack Johnson	1878–1946
Barney Oldfield	1878–1946
Miller Huggins	1879–1929
Willie Anderson	1880–1910
Christy Mathewson	1880–1925
Joe Tinker	1880–1948
Johnny Evers	1881–1947
Branch Rickey	1881–1965

Knute Rockne	1888–1931
Malcolm Campbell	1885–1948
Stanley Ketchel	1886–1910
Ty Cobb	1886–1961
Walter Johnson	1887–1946
Grover Cleveland Alexander	1887–1950
Willie Hoppe	1887–1959
Jim Thorpe	1888–1953
Tris Speaker	1888–1958
Earl Sande	1889–1968
Duke Kahanamoku	1890–1968
Casey Stengel	1891–1975
Walter Hagen	1892–1969
Bill Tilden	1893–1953
Francis Ouimet	1893–1967
Clifford Roberts	1893–1977
Babe Ruth	1895–1948
Benny Leonard	1896–1947
Rogers Hornsby	1896–1963
Paavo Nurmi	1897–1973
Eugene Tunney	1898–1978
Suzanne Lenglen	1899–1938
Buck Shaw	1899–1977
Joe Lapchick	1900–1970
Cal Hubbard	1900–1977
Adolph Rupp	1901–1977
Howie Morenz	1902–1937
Wilbur Shaw	1902–1954
Bobby Jones	1902–1971
Lou Gehrig	1903–1941
Mickey Cochrane	1903–1962
Ernie Nevers	1903–1976
Hirsch Jacobs	1904–1970
Wilmer Allison	1905–1977

Primo Carnera	1906–1967	Jackie Robinson	1919–1972
Jimmy Foxx	1907–1967	Rocky Marciano	1923–1969
Frank Leahy	1908–1973	Terry Sawchuck	1929–1970
Eric Shipton	1908–1977	Big Daddy	
Josh Gibson	1911–1947	Lipscomb	1931–1963
Sonja Henie	1912–1969	Maureen Connolly	1934–1969
Vince Lombardi	1913–1970	Jimmy Clark	1936–1968
Babe Didrikson			
Zaharias	1914–1956		

SUI GENERIS

Lady Godiva	c. 1010–1067	William Bonney	
Tomás de		("Billy the Kid")	1859–1881
Torquemada	c. 1420–1498	Lizzie Borden	1860–1927
Captain William		Butch Cassidy	1866–1937
Kidd	c. 1645–1701	Mata Hari	1876–1917
Giovanni Jacopo		Helen Keller	1880–1968
Casanova	1725–1798	Vidkun Quisling	1887–1945
Benedict Arnold	1741–1801	Hermann Göring	1893–1946
Captain William		Joseph Goebbels	1897–1945
Bligh	1754–1817	Amelia Earhart	1898–1937
Marie Duplessis	1824–1847	Alphonse ("Al")	
Lola Montez	c. 1818–1861	Capone	1899–1947
John Wilkes		Lavrenti Beria	1899–1953
Booth	1838–1865	Heinrich Himmler	1900–1945
Mattie Silks		Adolf Eichmann	1906–1962
("Queen of the		Joseph R.	
Red Lights")	1847–1929	McCarthy	1908–1957
Fannie Farmer	1857–1915	Lee Harvey	
		Oswald	1939–1963